Asymmetric Synthesis

Asymmetric Synthesis

Edited by

R. A. AITKEN
Department of Chemistry
University of St Andrews
UK

and

S. N. KILÉNYI
Sanofi Research
Bruxelles
Belgium

BLACKIE ACADEMIC & PROFESSIONAL
An Imprint of Chapman & Hall
London · Glasgow · New York · Tokyo · Melbourne · Madras

Published by Blackie Academic & Professional, an imprint of Chapman & Hall, Wester Cleddens Road, Bishopbriggs, Glasgow G64 2NZ

Chapman & Hall, 2–6 Boundary Row, London SE1 8HN, UK

Blackie Academic & Professional, Wester Cleddens Road, Bishopbriggs, Glasgow G64 2NZ, UK

Chapman & Hall, 29 West 35th Street, New York NY10001, USA

Chapman & Hall Japan, Thomson Publishing Japan, Hirakawacho Nemoto Building, 6F, 1-7-11 Hirakawa-cho, Chiyoda-ku, Tokyo 102, Japan

DA Book (Aust.) Pty Ltd, 648 Whitehorse Road, Mitcham 3132, Victoria, Australia

Chapman & Hall India, R. Seshadri, 32 Second Main Road, CIT East, Madras 600 035, India

First edition 1992

© 1992 Chapman & Hall

Printed in Great Britain at the University Press, Cambridge

ISBN 0 7514 0059 9

Apart from any fair dealing for the purposes of research or private study, or criticism or review, as permitted under the UK Copyright Designs and Patents Act, 1988, this publication may not be reproduced, stored, or transmitted, in any form or by any means, without the prior permission in writing of the publishers, or in the case of reprographic reproduction only in accordance with the terms of the licences issued by the Copyright Licensing Agency in the UK, or in accordance with the terms of the licences issued by the appropriate Reproduction Rights Organization outside the UK. Enquiries concerning reproduction outside the terms stated here should be sent to the publishers at the Glasgow address printed on this page.

The publisher makes no representation, express or implied, with regard to the accuracy of the information contained in this book and cannot accept any legal responsibility or liability for any errors or omissions that may be made.

A catalogue record for this book is available from the British Library

Library of Congress Cataloging-in-Publication data available

Foreword

Very few fields in science have shown such an acceleration from theory to practice as the field of asymmetric synthesis. When one considers that the phrase 'asymmetric synthesis' was just a mechanistic curiosity in 1965 with no one seriously believing that this could someday be an important part of molecular synthesis, the rate of progress has been quite remarkable. In just over twenty years the organic chemist has transformed this virtually unknown aspect of synthesis into a serious route to virtually every class of chiral organic compounds in greater than 90% enantiomeric purity. It is presently still enjoying significant growth and our understanding of many of the guiding principles makes it necessary to introduce the subject to students at a stage just after the initial introduction to traditional organic chemistry. It can now be safely stated that asymmetric synthesis may take its place among other, more traditional, methods for reaching enantiomerically pure compounds (resolution, microbiological processes, enzyme catalyzed reactions).

This monograph does an admirable job of bringing all the aspects of stereochemical principles and their application to a variety of asymmetric processes, to the young student of organic chemistry. Although not exhaustive in its coverage, it lays out the foundation in excellent form, with sufficient examples, so that the student can gain an appreciation of the progress made to date.

The authors, one of whom (Dr Aitken) has served as a postdoctoral fellow in my laboratory and has seen first hand the excitement of developing new asymmetric reactions, have done a fine job in outlining the principles, definitions, analytical techniques, and chemical behavior which surround this endeavor. The authors are to be congratulated for creating this monograph, for it will bring a new excitement to students in organic chemistry and it will hopefully point the way for them to pursue more advanced texts and literature compendia which will only serve to enhance their knowledge of one of the most important fields of modern chemistry.

Colorado State University A. I. Meyers
Fort Collins, Colorado, USA

Preface

This book has its origin in an eight lecture course on asymmetric synthesis given to final year B.Sc. students at the University of St Andrews since 1985. While there are many excellent monographs and reviews available covering specific topics, these are generally aimed at the research level, and none seemed entirely suitable as an introductory text for students new to the area. At the same time, the standard organic textbooks generally treat the subject in an outdated way, very briefly, or not at all. We have therefore attempted to fill this gap by providing an intermediate level book which assumes little more than an elementary knowledge of organic reactivity and stereochemical principles. We hope that it will prove useful as an introduction to the area for advanced undergraduate and postgraduate students in the UK and overseas, and also for chemists in industry who are not familiar with the area.

The first two chapters cover the aim and importance of asymmetric synthesis, as well as the sometimes complex terminology necessary for an understanding of the subject. The terminology associated with organic stereochemistry is a rapidly developing and often controversial area and we make no apology for trying to straighten out some common misconceptions here and provide the reader with a simple, up-to-date and self-consistent vocabulary for the rest of the book. Chapter 3 provides an account of the sophisticated methods now available to judge the degree of selectivity achieved in an asymmetric synthesis. In chapter 4 the general strategy of asymmetric synthesis is examined including the chiral starting materials available and a classification of the known methods. The heart of the book is chapters 5 and 6 in which we have tried to present and explain what we feel to be some of the most important asymmetric reactions. Here we faced an almost impossible task due to the sheer volume of material. Even concentrating on the best established and understood methods, there has only been space for a brief treatment of each and wherever possible we have included references to the most recent reviews so that the interested reader can get more information. We apologise in advance to anyone who finds that

PREFACE

their favourite method is not included. It is worth noting that the subject is still developing rapidly and the mechanistic details of some of the newer methods are not yet fully understood. In the final chapter some examples of total syntheses are presented to show the methods in action.

It is a pleasure for us to thank all the contributors for their valuable work. We must also thank Kit Finlay, Margaret Shand, Iona Hutchison and Lesley Arnott for typing the bulk of the manuscript and Shaun Mesher for drawing the structures for chapter 6.

R. A. Aitken
S. N. Kilényi

Contents

1 Chirality 1
R. A. AITKEN

1.1	The phenomenon of chirality	1
1.2	The biological significance of chirality: the need for asymmetric synthesis	2
1.3	The selective synthesis of enantiomers	5
1.4	The enantiomeric purity of natural products	7
1.5	The stereogenic unit and types of chiral compound	8
1.6	Centrally chiral compounds of carbon and silicon	9
1.7	Centrally chiral compounds of nitrogen and phosphorus	10
1.8	Centrally chiral compounds of sulphur	12
1.9	Axially chiral compounds	13
1.10	Chiral molecules with more than one stereogenic unit: diastereomers	15
1.11	The selective synthesis of diastereomers	18
1.12	Prochirality: enantotopic and diastereotopic groups	19
1.13	Absolute configuration	20
	Reference and notes	21

2 The description of stereochemistry 23
R. A. AITKEN

2.1	Compounds with one stereogenic centre	23
2.2	Axially chiral compounds	24
2.3	Compounds with more than one stereogenic unit	25
2.4	The topicity of enantioselective reactions	27
2.5	The relative topicity of diastereoselective reactions	28
2.6	Further points on correct terminology	30
	References and notes	31

3 Analytical methods: determination of enantiomeric purity 33
D. PARKER and R. J. TAYLOR

3.1	Polarimetric methods	33
3.2	Gas chromatography methods	36
3.3	Liquid chromatographic methods	39
3.4	NMR spectroscopy	42
	3.4.1 Chiral derivatising agents (CDAs)	44
	3.4.2 Chiral solvating agents (CSAs)	50
	3.4.3 Chiral lanthanide shift reagents (CLSRs)	57
3.5	Concluding remarks	60
	References and notes	62

4 Sources and strategies for the formation of chiral compounds 64
R. A. AITKEN and J. GOPAL

4.1	Chiral starting materials	64
	4.1.1 Amino acids and amino alcohols	64
	4.1.2 Hydroxy acids	66

CONTENTS

	4.1.3 Alkaloids and other amines	67
	4.1.4 Terpenes	68
	4.1.5 Carbohydrates	69
4.2	Methods for the formation of chiral compounds	70
	4.2.1 Use of naturally occurring chiral compounds as building blocks	70
	4.2.2 Resolution	71
	4.2.3 Methods of asymmetric synthesis	73
	4.2.4 Special methods	77
4.3	Mechanistic considerations	80
	References and notes	82

5 First- and second-generation methods: chiral starting materials and auxiliaries 83
S. N. KILENYI

5.1	Non-stereodifferentiating methods	83
	5.1.1 Classical resolution ((1R)-cis-permethric acid)	83
	5.1.2 Resolution using a chiral auxiliary	85
	5.1.3 Sulphoximines	87
	5.1.4 Methods using enantiomerically pure building blocks	90
5.2	First-generation methods	92
	5.2.1 Sugars	93
	5.2.2 Amino acids	96
	5.2.3 Terpenoids	96
	5.2.4 Hydroxy acids	98
5.3	Second-generation methods: nucleophiles bearing a chiral auxiliary	102
	5.3.1 General principles	102
	5.3.2 Chiral enolate and aza-enolate equivalents	102
	5.3.3 Asymmetric aldol reactions	109
	5.3.4 Asymmetric α-amino anions	118
	5.3.5 Chiral sulphoxides	119
5.4	Electrophiles bearing chiral auxiliaries	123
	5.4.1 Asymmetric Michael additions	123
	5.4.2 Chiral acetals	130
	5.4.3 Asymmetric additions to carbonyl compounds	132
5.5	Chiral auxiliaries in concerted reactions	135
	5.5.1 Diels-Alder cycloaddition	135
	5.5.2 The Claisen-Cope rearrangement	139
	References	140

6 Third- and fourth-generation methods: asymmetric reagents and catalysts 143
S. N. KILENYI and R. A. AITKEN

6.1	C—C bond-forming reactions	143
	6.1.1 Asymmetric alkylation	143
	6.1.2 Asymmetric Michael reaction	145
	6.1.3 Asymmetric nucleophilic additions to carbonyl compounds	149
	6.1.4 Asymmetric [2 + 2] cycloadditions	150
	6.1.5 Asymmetric Diels-Alder reaction	152
	6.1.6 Crotylboranes	153
	6.1.7 Asymmetric formation of alkene double bonds	155
6.2	Chiral acids and bases	156
6.3	Asymmetric oxidation methods	160
	6.3.1 Asymmetric epoxidation of alkenes	160
	6.3.2 Asymmetric oxidation of sulphides	162
	6.3.3 Asymmetric dihydroxylation	163
	6.3.4 Chiral oxaziridines and their uses	165

6.4	Asymmetric reduction, double bond isomerisation and hydroboration	167
	6.4.1 Catalytic hydrogenation with chiral transition metal complexes	167
	6.4.2 Asymmetric double bond isomerisation	173
	6.4.3 Asymmetric hydroboration of alkenes	174
	6.4.4 Asymmetric reduction using chiral boranes and borohydrides	176
	6.4.5 Chirally modified $LiAlH_4$	179
6.5	Enzymatic and microbial methods	181
	6.5.1 Enzymatic reduction	181
	6.5.2 Enantioselective microbial oxidations	185
	6.5.3 Esterases and lipases	188
References		189

7 Asymmetric total synthesis 192
S. N. KILENYI

7.1	First-generation methods	192
	7.1.1 (R)-(+)-Frontalin	192
	7.1.2 (−)-Prostaglandin E_1	193
	7.1.3 The Merck synthesis of (+)-thienamycin	196
7.2	Second-generation methods	200
	7.2.1 Prelog–Djerassi lactone	200
	7.2.2 (R)-Muscone	203
	7.2.3 (−)-Podophyllotoxin	203
	7.2.4 (−)-Steganone	206
	7.2.5 (+)-α-Allokainic acid	213
	7.2.6 Leiobunum defence secretion	215
7.3	Third-generation methods	216
	7.3.1 Carbocyclic iododeoxyuridine	216
7.4	Fourth-generation methods	218
	7.4.1 A tricyclic analogue of indacrinone	218
	7.4.2 (S)-(−)-Propranolol	220
	7.4.3 Takasago (−)-menthol synthesis	222
	7.4.4 (−)-Prostaglandin E_1	222
	7.4.5 Compactin and mevinolin analogues	224
References		227

Index 231

Contributors

Dr R. Alan Aitken Department of Chemistry, University of St Andrews, Purdie Building, St Andrews, Fife KY16 9ST, UK

Dr Jayalakshmi Gopal Department of Chemistry, University of St Andrews, Purdie Building, St Andrews, Fife KY16 9ST, UK

Dr S. Nicholas Kilényi Sanofi Research, Avenue de Bejar 1, Bruxelles, B-1120, Belgium

Dr David Parker Department of Chemistry, University of Durham, South Road, Durham DH1 3LE, UK

Dr Richard J. Taylor Department of Chemistry, University of Durham, South Road, Durham DH1 3LE, UK

1 Chirality

R. A. AITKEN

Asymmetric synthesis involves the formation of chiral molecules. In this chapter the nature and significance of chirality are described. This is followed by a survey of the major types of chiral organic molecules and the terminology used to describe them.

1.1 The phenomenon of chirality

Chirality is a fundamental symmetry property of three-dimensional objects. An object is said to be chiral if it cannot be superimposed upon its mirror image. In a chemical context, chirality is applied to the three-dimensional structure of molecules. Many compounds may be obtained in two different forms in which the molecular structures are constitutionally identical but differ in the three-dimensional arrangement of atoms such that they are related as mirror images. In such a case the two possible forms are called **enantiomers** and are said to be **enantiomeric** with each other. To take a simple example, the amino acid alanine can be obtained in two forms (**1**) and (**2**) which are clearly related as mirror images.

Enantiomers have identical chemical and physical properties in the absence of an external chiral influence. This means, for example that (**1**) and (**2**) have the same melting point, solubility, chromatographic retention time and IR and NMR spectra as each other. If (**1**) and (**2**) are mixed, however, the resulting sample will have different physical properties, such as melting point and solubility, but the chemical based properties, such as chromatographic and spectroscopic behaviour will be unchanged. This has an important consequence if we want to determine the proportion of the two enantiomers in a mixture: the normal chromatographic and spectroscopic methods of analysis must be modified to *introduce an external chiral influence*. Only then will the enantiomers behave differently from each other and analysis be possible. This is discussed in detail in chapter 3.

There is one property in which enantiomers do differ and that is the direction in which they rotate the plane of plane-polarised light. This phenomenon of **optical activity** provides the basis for the nomenclature of enantiomers. Thus (**1**), which rotates the plane of plane-polarised light in a clockwise direction ($[\alpha]_D = +14.6°$ ($c = 1$, 5M HCl)),[1] is denoted (+)-alanine while the enantiomer (**2**) which has an equal and opposite rotation under the same conditions ($[\alpha]_D = -14.6°$) is denoted (–)-alanine. Since the rotations due to the individual enantiomers in a mixture are additive, to a first approximation, the net measured rotation may be used as a guide to the enantiomeric composition.

1.2 The biological significance of chirality: the need for asymmetric synthesis

If the only difference between enantiomers was in their rotation of plane-polarised light, the whole area of asymmetric synthesis would be relegated to little more than a scientific curiosity. That this is not so is because the world around us is chiral and most of the important building-blocks which make up the biological macromolecules of living systems do so in one enantiomeric form only. When, therefore, a biologically active chiral compound, such as a drug, interacts with its receptor site which is chiral, it should come as no surprise that the two enantiomers of the drug interact differently and may lead to different effects.

A good example is the drug thalidomide (**3**) for which both enantiomers have the desired sedative effect but only the (–)-enantiomer (shown here) causes foetal deformities. Unfortunately the drug was used clinically as an equal mixture of the two enantiomers but even if the pure (+)-enantiomer had been used problems would have arisen since the two interconvert under physiological conditions.

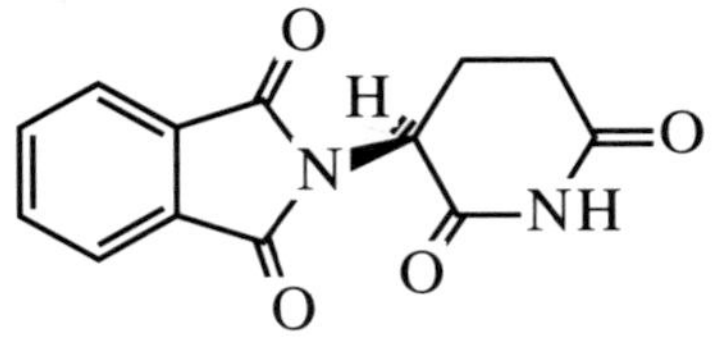

(3) 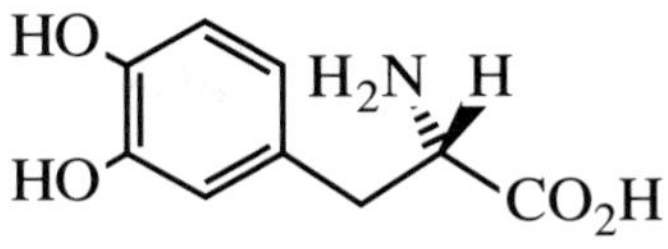   (4)

A more interesting situation occurs in the case of DOPA (4) used in the treatment of Parkinson's disease. The active drug is the achiral compound dopamine formed from (4) by decarboxylation, but this cannot cross the 'blood-brain barrier' to reach the required site of action. The 'prodrug' (4) can and is then decarboxylated by the enzyme dopamine decarboxylase. The enzyme, however, is specific and only decarboxylates the (−)-enantiomer of (4) (the form shown here). It is therefore essential to administer DOPA as the pure (−)-enantiomer otherwise there would be a dangerous build up of (+)-(4) in the body which could not be metabolised by the enzymes present. The asymmetric synthesis of (4), which is carried out on an industrial scale, is described in section 6.4.1.

For many chiral compounds the two enantiomers have quite distinct biological activities. (−)-Propranolol (5) was introduced in

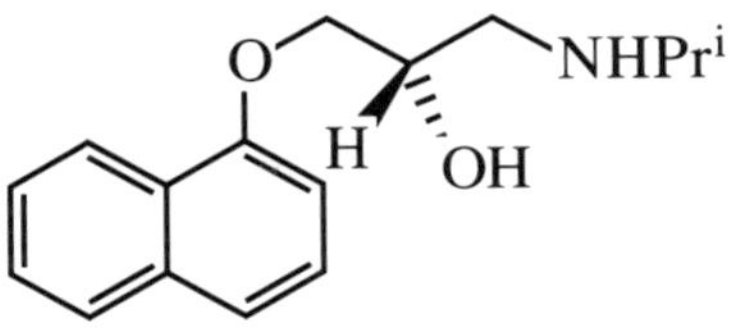

(5) 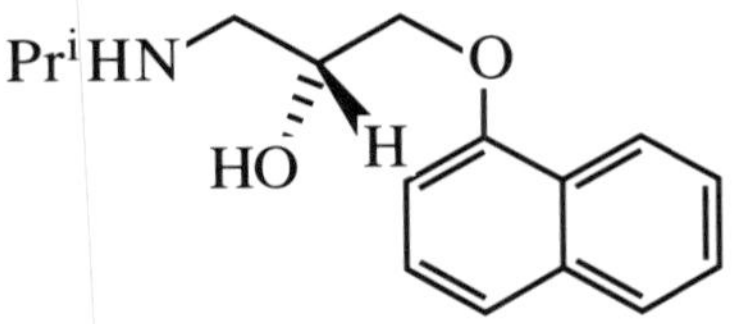   (6)

the 1960s as a β-blocker for the treatment of heart disease, but the (+)-enantiomer (6) acts as a contraceptive so enantiomeric purity was obviously essential for clinical use. The asymmetric synthesis of (5) is described in section 7.4.2.

The alkaloid (−)-levorphanol (7) is a powerful narcotic analgaesic with an activity 5–6 times stronger than morphine. Its enantiomer, (+)-dextrorphan (8), is totally devoid of this activity, but is active as a cough supressant and is marketed for this purpose as its methyl ether (dextromethorphan).

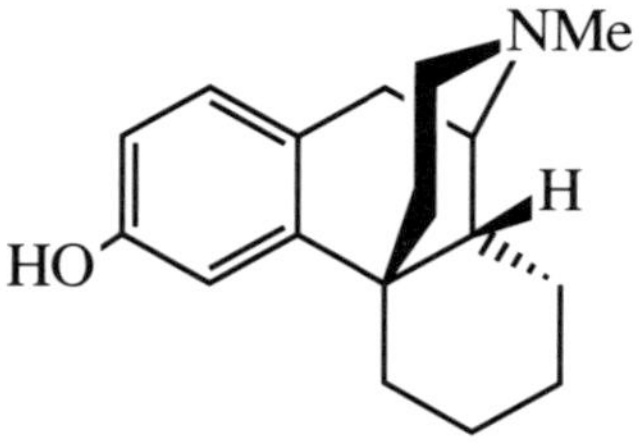

(7) 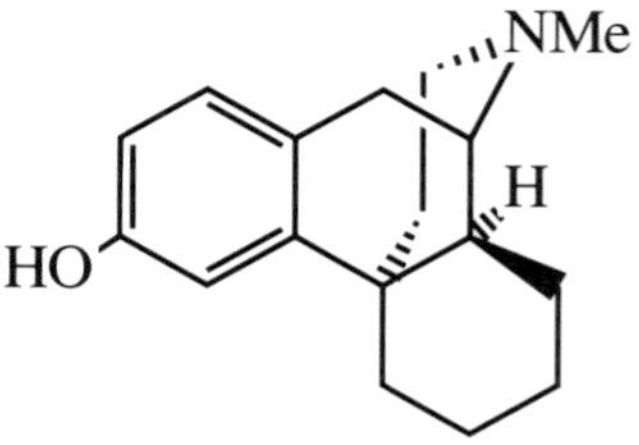 (8)

In the last example there is more than one tetrahedral centre giving rise to stereoisomers (**stereogenic centre**) and this brings us to a most important point in favour of asymmetric synthesis: for more complex chiral molecules there are not only two possible stereoisomeric forms (enantiomers) but anywhere up to 2^n where n is the number of **stereogenic units** (see section 1.5). While a conventional synthesis followed by separation of the two enantiomers (for example by resolution) is feasible when there is only one stereogenic unit, in chiral compounds with several stereogenic units the desired product, with correct stereochemistry in each unit, can in general only be prepared using asymmetric synthesis.

This is demonstrated by the case of the artificial sweetener aspartame. Only the stereoisomer (**9**) is sweet and the other three (**10–12**) are all slightly bitter and must be avoided in the manufacturing process.

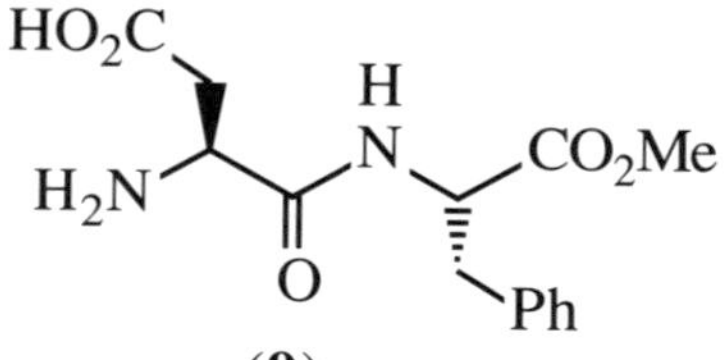

(9) 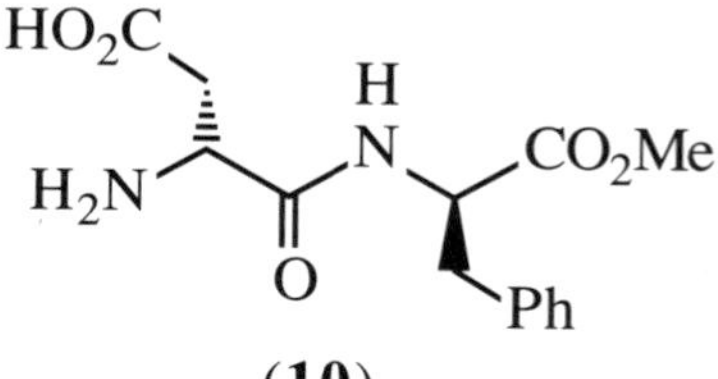 (10) 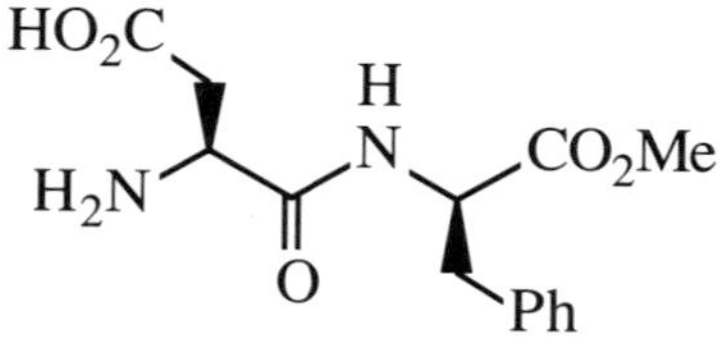 (11) 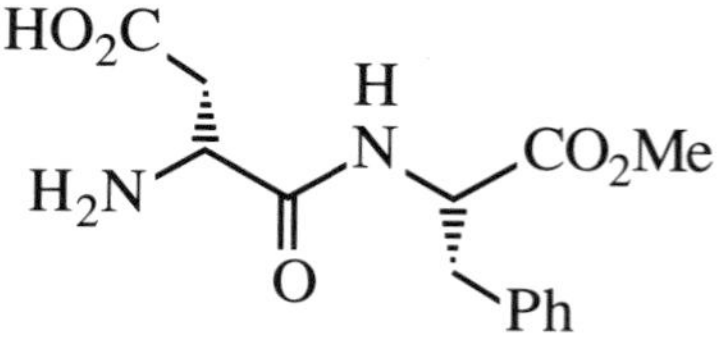 (12)

CHIRALITY

Even when the other stereoisomers are inert it may still be desirable to synthesise and use the active one in pure form. The first reason for this is economic. The formation of inert isomers represents a waste of starting materials and resources. Even an expensive asymmetric method may be justified if it gives exclusively the active stereoisomer. The second reason, which is becoming more and more important, might be termed environmental. Although inactive stereoisomers may appear to be inert in the short term, unless they are rapidly and safely biodegraded there is a risk of long-term side effects. Thus it is clearly undesirable for an active pharmaceutical product to be administered along with several inactive isomers which do no good and might be harmful. Similarly, if only the active isomer of chiral agrochemicals such as insecticides is used the environmental impact is minimised. For the pyrethrin insecticide deltamethrin (**13**) there are eight possible stereoisomers but only the one shown is active and photostable.

(**13**)

Although it would be desirable to use all biologically active compounds as the pure active stereoisomer, this would be prohibitively expensive in some cases. However, it is now common practice to at least synthesise and evaluate all the possible isomers of a new product before it is put into use.

1.3 The selective synthesis of enantiomers

An asymmetric synthesis may be defined as a synthesis in which an achiral unit in an ensemble of substrate molecules is converted to a chiral unit such that the possible stereoisomers are formed in unequal amounts.

In the simplest case an achiral substrate is converted to an unequal mixture of the two enantiomers of a chiral product containing only one stereogenic unit. The aim is obviously to achieve the highest possible proportion of the desired enantiomer: to maximise the **enantioselectivity**. The most commonly used measure of the degree of enantioselectivity achieved is the **enantiomeric excess (e.e.)**. This is defined as the proportion of the major enantiomer less that of the minor enantiomer and is commonly expressed as a percentage. Thus, for example, if the reduction of acetophenone is carried out asymmetrically to give the enantiomeric alcohols (**14**) and (**15**) in a ratio of 90:10, then the e.e. of the process is 80%. Similarly an e.e. of 90% would correspond to an enantiomer ratio of 95:5. The reason for using the enantiomeric *excess* rather than the enantiomer ratio is that in almost all cases[2] it corresponds directly with the optical purity. Thus in the example above (**14**) has an optical rotation of –120° and (**15**) of +120°. A sample of 80% e.e. which contains 90% of (**14**) and 10% of (**15**) will have a net optical rotation of (0.9 × –120°) +(0.1 × +120°) = –96° which is 80% of the value for the pure major enantiomer.

Thus for any sample of a compound for which the optical rotation of the pure enantiomer is known, the e.e. can be determined directly from the observed rotation.

Special mention should be made of the two extreme values of e.e. An e.e. of 100% corresponds to an **enantiomerically pure compound** (the term *homochiral* which is also sometimes used is not favoured). A reaction which gives a product of 100% e.e. is **enantiospecific**. Since this represents an ideal situation which is rarely attainable in practice, the term **enantioselective** should generally be used. An e.e. of 0% corresponds to a 1:1 mixture of enantiomers known as a **racemic mixture** or **racemate** (this is denoted by the prefix (±)-). The process by which the stereogenic unit in a chiral compound is destroyed and then reformed with random stereochemistry leading to a fall in the e.e., eventually to zero, is described as **racemisation**. Note that even in a racemic compound each molecule is individually chiral and this has led some authors to use the term *chiral non-racemic* for the product from an enantioselective reaction.

Finally in this section it should be noted that the solubility of a pure enantiomer and the racemate are not necessarily equal. For a racemic solid compound recrystallisation can give racemic crystals (containing equal proportions of both enantiomers) or individual crystals of each enantiomer which can sometimes be separated. In rare cases there can even be spontaneous crystallisation of one pure enantiomer although this should theoretically happen equally often to give the opposite enantiomer. As soon as we move to solid compounds with a significant e.e. however, it is quite common for the major enantiomer and the racemate to be separable by crystallisation. This phenomenon of **enantiomeric enrichment** is extremely useful since in many cases solid products of 60–80% e.e. from asymmetric synthesis can be enriched to over 95% e.e. by one or two recrystallisations.

1.4 The enantiometric purity of natural products

Until the development of the direct, non-polarimetric methods for the determination of e.e. which began in the 1960s, there was no accurate way of knowing the enantiomeric composition of the many naturally occurring chiral compounds. It was generally assumed that the highest optical rotation achieved after repeated attempts at enantiomeric enrichment, most commonly by *resolution* (see section 1.10), corresponded to the enantiomerically pure material. If samples of a chiral compound from several different routes had the same maximum rotation this further strengthened the argument. Against this background it was commonly believed that all naturally occurring chiral compounds were enantiomerically pure. In fact we now know that, while this is still generally true, there are many exceptions. Nature can and does synthesise many important chiral compounds in either enantiomeric form. The simple thietane (**16**) is a component of the anal scent secretion of the stoat, *Mustela arminea*, and samples from this source showed a rotation $[\alpha]_D = -40°$. However, a sample of the pure (–)-enantiomer of (**16**) obtained by asymmetric synthesis had a value of $[\alpha]_D = -147.5°$, showing that the natural product has an e.e. of 26% and consists of a mixture of 63% of (**16**) with 37% of the opposite enantiomer. Many recent asymmetric total syntheses of natural products have given products with slightly greater optical rotations than the natural materials which had previously been thought to be enantiomerically pure.

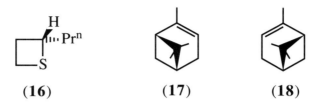

(16)   (17)   (18)

A second example of more practical importance is the terpene (+)-α-pinene (**17**). This chiral bicyclic alkene is of great value for the formation of chiral boron reagents (see section 6.4.3) but unfortunately the commercially available material obtained from pine resin has an e.e. of only 92% and contains 4% of the (−)-enantiomer (**18**). This limited the potential of the boron reagents until the recent development of a method of enantiomeric enrichment.

Despite the note of caution introduced in this section, nature does provide a large number of compounds in enantiomerically pure form and, as we shall see in chapter 4, these provide the basis for all asymmetric synthesis.

1.5 The stereogenic unit and types of chiral compound

A unit within a molecule which gives rise to the existence of stereoisomers is called a **stereogenic unit**. The chirality of most chiral molecules is associated with the presence of one or more stereogenic units although it is important to note that *the presence of a stereogenic unit is not in itself a sufficient condition for chirality*. This remains that the molecule should not be superimposable on its mirror image.

Simple chiral molecules can be classified into three types according to the type of stereogenic unit present: *central, axial* and *planar*. A *centrally chiral* molecule is chiral by virtue of the arrangement of atoms or groups about a **stereogenic centre**. The most familiar example is a tetrahedral molecule of type (**19**) and this is the most common class of chiral molecule. An *axially chiral* molecule is chiral by virtue of the arrangement of atoms or groups about a **stereogenic axis**. An example is provided by the biaryl (**20**). This class of chiral compound occurs quite commonly. The final type: *planar chirality*, in which the chirality is due to the arrangement of atoms or groups with respect to a **stereogenic plane** is illustrated by *E*-cyclooctene (**21**). This class of chiral compounds is quite rare and will not be considered further in this book.

1.6 Centrally chiral compounds of carbon and silicon

The vast majority of all chiral compounds are centrally chiral and possess one or more tetrahedral stereogenic carbon centres. The compounds (**1**)–(**18**) mentioned earlier in this chapter are all of this type. Some additional examples of special interest are mentioned here. One of the simplest possible chiral compounds is (**22**) which was obtained in enantiomerically pure form only recently and shows a small rotation of $[\alpha]_D = +1.6°$ for the enantiomer shown. This brings us to an interesting feature of optical activity: the size of the rotation depends to some extent on the presence of a chromophore to interact with the incident light. In particular, chiral compounds which are completely non-polarised, such as saturated hydrocarbons, may have immeasurably small optical rotations. For example, the rotation of enantiomerically pure hydrocarbon (**23**) which was prepared in 1959 by Wynberg, was predicted to be ~0.00002° but (not surprisingly) could not be detected on any available polarimeter.

Small rotations are also observed for another interesting group of chiral compounds: those which are chiral due to the presence of isotopes. The chiral benzyl alcohol (**24**) and the diphenylmethanol (**25**) have both been obtained as pure enantiomers and show rotations of +0.7° and −1.0° respectively. Even the acetic acid (**26**) chiral due

(**24**) Ph−C(H)(D)−OH

(**25**) C_6H_5−C(H)(OH)−C_6D_5

(**26**) T,D,H−C−CO_2H

to the presence of all three isotopes of hydrogen has been prepared by Arigoni and used to great effect in the elucidation of enzymic stereoselectivity in biosynthesis.

Spiro compounds, in which the stereogenic centre is at the fusion of two rings, can also be chiral, as in the example of the olive-fly pheromone (**27**).

(**27**) enantiomeric with

Silicon compounds which have an asymmetric arrangement of atoms around a tetrahedral silicon centre can, of course, also be chiral. There has been little work on chiral silicon compounds to date but a comprehensive review of this is available.[3]

1.7 Centrally chiral compounds of nitrogen and phosphorus

Unsymmetrically substituted amines (**28**) are potentially chiral because of the pseudo-tetrahedral arrangement of the three groups and the lone pair. Under normal conditions, however, there is rapid inversion at the nitrogen centre (racemisation) which prevents the separation of enantiomers. Only in special cases where the nitrogen is in a small ring with electron-withdrawing groups present have compounds been obtained which are chiral only because of the presence of an amine nitrogen. The examples (**29**) and (**30**) have rotations of +75° and −284° respectively.

As soon as the lone pair is 'fixed' by bonding to an electrophile, inversion is prevented and therefore both the quaternary ammonium

(**28**) R¹−N(··)(R³)−R²

(**29**) MeO₂C, MeO₂C, O−N(··)−Buᵗ

(**30**) Ph, Ph−N(··)−Cl

(**31**) R¹−N⁺(R⁴)(R³)−R²

(**32**) R¹−N⁺(O⁻)(R³)−R²

(**33**) R¹−P(··)(R³)−R²

salts (**31**) and tertiary amine N-oxides (**32**) are potentially chiral. The chiral compounds of nitrogen have been reviewed.[4]

In contrast to the amines, inversion of configuration in phosphines is generally negligible at room temperature and the phosphines (**33**) are a well known class of chiral compounds which are of particular value as ligands in transition metal catalysed asymmetric synthesis (see section 6.4.1). The following example shows the preparation of (+)-CAMP and (−)-DIPAMP. The racemic starting material (**34**) is first *resolved* (see section 1.10) by reaction with enantiomerically pure (−)-menthol to give a mixture of the sulphinate esters (**35**) and (**36**). These are separated to give the pure stereoisomer (**36**) which then reacts with *o*-anisylmagnesium bromide, with inversion of configuration, to afford the enantiomerically pure phosphine oxide (**37**). Selective hydrogenation of the phenyl ring in (**37**) followed by deoxygenation with inversion of configuration at phosphorus gives (+)- CAMP, while oxidative coupling through the methyl group and deoxygenation, again with inversion, gives (−)-DIAMP.

As well as illustrating the synthesis of chiral phosphines this sequence introduces two other types of chiral phosphorus compound: the phosphinate ester (**36**) and the phosphine oxide (**37**). In fact any tetracoordinate phosphorus compound with four different groups attached to the tetrahedral centre is chiral. The chiral compounds of phosphorus have been reviewed.[5]

1.8 Centrally chiral compounds of sulphur

The most important class of chiral sulphur compounds is the sulphoxides. As a result of the presence of the lone pair, the stereochemistry at sulphur is pyramidal and molecules of the general type (**38**) and (**39**) are opposite enantiomers. The sulphoxides are of considerable importance in asymmetric synthesis (see chapter 5) and, as for the phosphines, are generally obtained in enantiomerically pure form by resolution.

If the lone pair in (**38**) or (**39**) is replaced by a double bond to oxygen the resulting sulphone is obviously achiral but if it is instead replaced by a double bond to nitrogen, we get another type of chiral sulphur compound, the sulphoximines (**40**) which have also been used for asymmetric synthesis (see section 5.1.3). The sulphonium salts (**41**) are also potentially chiral. A review of the chiral compounds of sulphur is available.[6]

1.9 Axially chiral compounds

The two most important classes of axially chiral compounds which will be considered here are the allenes and hindered biaryls.

At first sight an allene (**42**) might be considered as a 'stretched out' tetrahedron. As in the tetrahedron the planes aCb and cCd are at right angles to each other, and the result is that allenes of the general type (**42**) are chiral. This is to be contrasted with the alkenes (**43**) where there is E–Z-isomerism. In fact these are just the first members of a whole series of compounds, the cumulenes, in which those with an odd number of cumulated carbon atoms are potentially chiral while those with an even number of carbon atoms show E–Z-isomerism. The move from a tetrahedral molecule to an allene has a further important consequence: the C_3 rotational symmetry element of the tetrahedral shape is removed and this means that it is no longer necessary for all four groups a–d to be different. It is sufficient for the pair at each end to be different from each other. Examples are provided by the enantiomerically pure allenes (**44**) and (**45**) which have rotations of –314° and –124° respectively. The allene function also turns up occasionally in natural products and the pheromone of the male dried

bark beetle, *Acanthoscelides obtectus* is found with an e.e. of 74% in favour of the enantiomer (**46**) which has $[\alpha]_D = -176°$. Chiral allenes have been reviewed.[7]

A related type of axially chiral molecule is exemplified by the 4-alkyl alkylidenecyclohexane (**47**) which might be viewed as an allene with one double bond expanded out into a ring. Compounds

of this type are unusual in that (**47**) can be converted to the opposite enantiomer (**48**) either by inversion at the tetrahedral centre or by *E–Z* isomerisation of the alkene function. The asymmetric synthesis of such a compound is described in sections 5.3.5 and 6.1.7.

The hindered biaryls are examples of a different type of chirality which arises due to hindered rotation about a C–C single bond. As long as the *ortho*-substituents in a compound such as (**49**) are large enough that they cannot pass each other, the compound can exist in two forms, (**49**) and the enantiomer (**50**) which cannot interconvert.

CHIRALITY 15

(49)   (50)   (51)   (52)

In fact the rings are at right angles to each other and the groups a–d take up a tetrahedral arrangement in space. As for the allenes, it is no longer required that all four groups be different from each other. An example is provided by the dinitro-diacid (**51**). Chiral compounds like this, in which the enantiomers can in principle be interconverted without breaking any covalent bonds, are said to exhibit **conformational chirality** or **atropisomerism** and the enantiomers are called **atropisomers**. The 2,2'-disubstituted binaphthyls (**52**) provide another example of atropisomerism due to the steric clash of the hydrogen atoms on position 8 of each naphthalene nucleus. The asymmetric synthesis of chiral biaryls is covered in section 5.4.1 and an example of the synthesis of a natural product containing a chiral biaryl stereogenic unit is given in section 7.2.4.

1.10 Chiral molecules with more than one stereogenic unit: diastereomers

As mentioned earlier there are only two possible stereoisomers of a chiral compound containing one stereogenic unit: the (+)- and (−)-enantiomers. If there are n stereogenic units there can be anywhere up to 2^n stereoisomers. In comparing any two of these stereoisomers two possibilities arise: either they are mirror images of each other, in which case they are enantiomers, or they are not in which case they are called **diastereomers**. An example is provided by the four possible stereoisomers of the amino acid threonine.

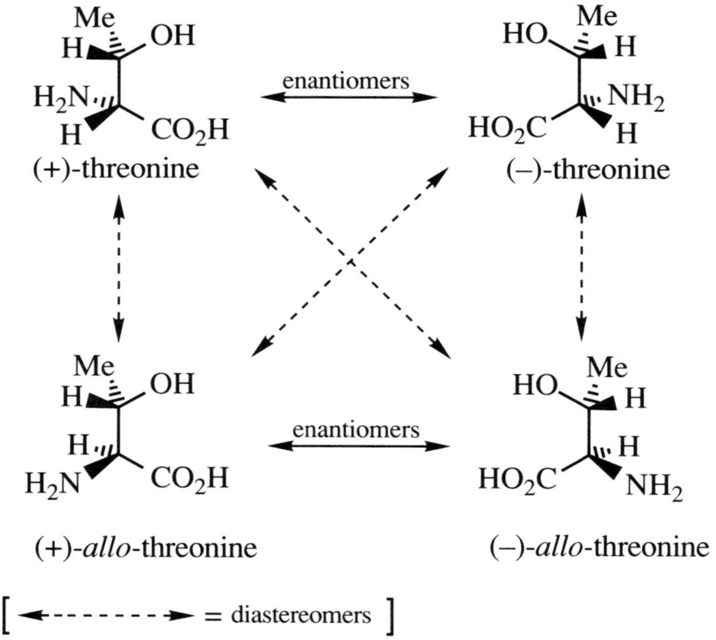

ASYMMETRIC SYNTHESIS

[Diagram showing four stereoisomers of threonine: (+)-threonine and (−)-threonine (enantiomers, top row); (+)-*allo*-threonine and (−)-*allo*-threonine (enantiomers, bottom row); with dashed arrows indicating diastereomeric relationships between the pairs.]

[⟵ - - - - - - ⟶ = diastereomers]

While compounds which are enantiomers have identical chemical and physical properties and equal and opposite optical rotation, compounds which are **diastereomeric** with each other can have completely different chemical and physical properties and optical rotations. This feature provides the basis for the *resolution* of chiral compounds. In this procedure a racemic mixture is derivatised by reaction with an enantiomerically pure compound which leads to a mixture of two diastereomers. These can be separated by crystallisation or chromatography as a result of their different properties and the pure enantiomers of the starting compound obtained by cleavage of each diastereomer separately. In the synthesis of the chiral phosphines (section 1.7) the phosphine oxide (**37**) is obtained by resolution of the starting material (**34**) *via* the diastereomeric esters (**35**) and (**36**). A further important application of the differing properties of diastereomers is in the determination of e.e. by derivatisation with an enantiomerically pure compound followed by chromatographic or NMR analysis (see section 3.4.1).

While enantiomers normally bear the same name and are differentiated by the prefixes (+)- and (−)-, diastereomers may have different chemical names, as in the case of threonine and *allo*-threonine above, although within each enantiomeric pair the enantiomers are still denoted (+)- and (−)- and have the same name. To take a well known example, there are a total of 16 possible stereoisomers of the hexoses $HOCH_2$-$(CHOH)_4$-CHO with four stereogenic centres. Each of the eight enantiomeric pairs has a different name, giving (+)- and (−)-glucose, (+)- and (−)-galactose, (+)- and (−)-mannose, etc.

A further example of the diastereomeric relationship is provided by the naturally occurring alkaloids cinchonine and cinchonidine and their methoxy derivatives quinidine and quinine which all occur together in the bark of the *cinchona* tree. Note that while the stereogenic centres at C-8 and C-9 have the opposite configuration, the three

R=H   (−)-cinchonidine   (+)-cinchonine
R=OMe   (−)-quinine   (+)-quinidine

stereogenic centres in the bicyclic quinuclidine part remain the same. Only the presence of the vinyl group prevents these compounds from being enantiomeric.

The diastereomeric alkaloids (−)-ancistrocladine (**53**), and (+)-hamatine (**54**), which are both obtained from the plant *Ancistrocladus hamatus* have the opposite configuration in the axial stereogenic unit (atropisomers) but the same configuration in the two stereogenic centres of the tetrahydroisoquinoline ring. Aromatisation would again give an enantiomeric pair.

(**53**)   (**54**)

1.11 The selective synthesis of diastereomers

Perhaps the most common case in which an asymmetric reaction leads to a pair of diastereomers is when one of the reactants is chiral. If we consider reduction of the carbonyl group in (**55**) for example, the two possible products are (–)-threonine (**56**) and (+)-*allo*-threonine (**57**) which, as we have already seen, are diastereomeric with each other.

![Reduction of 55 to give 56 and 57]

Several different measures of the **diastereoselectivity** can be given. Just as for e.e., we can define the **diastereomeric excess (d.e.)** of a reaction as the proportion of the major diastereomer produced less that of the minor one. In examples such as the one here, where one new stereogenic unit is formed in a diastereoselective reaction, this is the preferred measure of selectivity. While it does not have the same correlation with optical purity as e.e., it does have the advantage that if the original stereogenic unit(s) are removed, as, for example, by removal of the chiral auxiliary in a second-generation method (see chapter 5), the e.e. of the final product correlates directly with the d.e. of the initial product. Thus if the mixture of (**56**) and (**57**) was decarboxylated, the e.e. of the resulting amino alcohol would be equal to the d.e. of (**56**)/(**57**).

The most useful alternative measure of diastereoselectivity is the **diastereomer ratio (d.r.)** which, in contrast to d.e., is expressed not as a percentage but as a ratio. To take a simple example, if reduction of (**55**) gives (**56**) and (**57**) in a ratio of 90:10, the d.e. is 80% and the d.r. 9:1. For reactions in which two new stereogenic units are formed the diastereoselectivity is best expressed as the diastereomer ratio (see below). We do not favour the quantity '% d.s.' used by some authors.[8] The terms **diastereoselective** and **diastereospecific** are used in an analogous way to enantioselective and enantiospecific. Note that a **diastereomerically pure** compound refers only to a pure enantiomeric pair of stereoisomers and gives no information on the ratio of the enantiomers present (see below).

A slightly more complex case of diastereoselectivity arises in reactions in which two stereogenic units are formed simultaneously. Suppose, for example, the aldol reaction of benzaldehyde and ethyl

OH Ph—⟨—CO₂Et Me (**58**) ratio 68	OH Ph—⟨—CO₂Et Me (**59**) 12	OH Ph—⟨—CO₂Et Me (**60**) 15	OH Ph—⟨—CO₂Et Me (**61**) 5

propionate can be carried out asymmetrically to give the four possible stereoisomeric products (**58–61**) in the proportions shown. The diastereomeric excess is given by the proportion of the major diastereomer [the enantiomeric pair (**58**) and (**59**)] less that of the minor diastereomer [(**60**) and (**61**)] and thus this reaction has 60% d.e. However, since in a case like this the d.e. has no special significance, it is preferable to quote the diastereomer ratio, i.e. 4:1. For each diastereomer the e.e. can also be given: 70% e.e. for the major diastereomer and 50% e.e. for the minor diastereomer. Of course it is usually only the e.e. for the major diastereomer that is of interest.

1.12 Prochirality: enantiotopic and diastereotopic groups

If two atoms or groups in an achiral molecule differ in that reaction of one as opposed to the other leads to enantiomeric products, they are said to be **enantiotopic** and the whole molecule is described as being **prochiral**. By far the most common situation in which this applies is when two of the four groups joined to a tetrahedral centre are identical. Thus the benzylic hydrogen atoms in benzyl alcohol (**62**) are enantiotopic as are the two oxygens in an unsymmetrical sulphone (**63**), and the two methyls of an isopropyl group (**64**).

OH Ph—⟨⟩—H H (**62**)	O ‖ S—R¹ O R² (**63**)	H ⟨⟩—CH₃ R CH₃ (**64**)

20   ASYMMETRIC SYNTHESIS

If one or more stereogenic units are present elsewhere in the molecule then reaction of one group as opposed to the other leads to diastereomeric products and the groups are said to be **diastereotopic**. The two isopropyl methyl groups in (+)-valine (**65**) are thus diastereotopic as are both the ring and benzylic CH_2 hydrogens in the thiazoline (**66**). It is worth remembering that these groups are still diastereotopic even in the racemic compounds since they are diastereotopic in each enantiomer. An important feature of diastereotopic groups is that they may be *magnetically non-equivalent* and so give separate NMR signals.

(**65**)   (**66**)

This terminology can also be applied to the faces of trigonal planar groups such as carbonyl. Thus the two faces of acetophenone are enantiotopic as seen by the fact that reaction of hydride ion at one as opposed to the other gives enantiomeric products (**14**) and (**15**) (section 1.3). The faces of the carbonyl group in (**55**) are diastereotopic because reduction on one or other gives diastereomers (**56**) and (**57**) (section 1.11).

1.13 Absolute configuration

Throughout this chapter we have written three-dimensional structures for particular stereoisomers and it is important to know how these have been determined. How, for example, do we know that (+)-alanine has the absolute configuration (**1**) and not (**2**)? The answer is that until 1951 this was not known and the three-dimensional structure of stereoisomers was shown according to a *convention* introduced in the last century by Emil Fischer. According to this convention it was assumed that (+)-glyceraldehyde had the three-dimensional structure (**67**). Once this assumption had been made a self-consistent system of conventional representations could be applied to many other chiral compounds by chemical correlation with (**67**). In 1951 a group led by Bijvoet in Utrecht determined the absolute configuration of the sodium rubidium salt of (+)-tartaric acid to be (**68**) by a special X-ray diffraction technique.[9] Fortunately this proved that Fischer's

(**67**)   (**68**)

convention did in fact correspond with reality and immediately gave the true absolute configurations of many thousands of chiral compounds whose relationship to (+)-glyceraldehyde had already been determined. For this reason it could be argued that the determination of the three-dimensional structure of (**68**) was the most important X-ray diffraction study ever performed. The determination of absolute configuration by X-ray techniques is now somewhat easier but is still not trivial. The absolute configuration of new chiral compounds can **only** be determined either by correlation with compounds of known configuration which can be traced back ultimately to an X-ray study or, if this is not possible, by making a new X-ray determination.

References and notes

1. A detailed explanation of this notation and the measurement of optical rotation is given in section 3.1
2. A few examples are known in which the optical purity and the enantiomeric excess are not equivalent. It is in any case generally advisable to determine e.e. by an independent method in addition to polarimetry (see chapter 3).
3. C.A. Maryanoff and B.E. Maryanoff in *Asymmetric Synthesis*, Vol. 4, J.D. Morrison and J.W. Scott, eds, Academic Press, Orlando, 1984, Chapter 5.
4. F.D. Davis and R.H. Jenkins in reference [3], Chapter 4.
5. D.R. Valentine in reference [3], Chapter 3.
6. M.R. Barbachyn and C.R. Johnson in reference [3], Chapter 2.
7. W. Runge in *The Chemistry of the Allenes*, Vol. 2, S.R. Landor, ed., Academic Press, 1982, Chapter 6.
8. The 'percentage diastereoselectivity' (% d.s.), which is simply the percentage of the major diastereomer formed, seems to us a most unsatisfactory term, since it does not correspond with the normal concept of selectivity. In particular, a completely non-selective reaction such as the formation of equal proportions of (**56**) and (**57**) from (**55**) would be described as having 50% d.s.
9. J.M. Bijvoet, A.F. Peerdeman and A.J. Bommel, *Nature (London)*, 1951, **168**, 271.

2 The description of stereochemistry
R. A. AITKEN

In this chapter the systems of nomenclature used to describe the absolute configuration of enantiomers and diastereomers as well as the possible modes of attack in enantioselective and diastereoselective reactions are explained.

2.1 Compounds with one stereogenic centre

We have already seen that enantiomers can be denoted, according to their sign of optical rotation, as (+) and (−). For an understanding of asymmetric synthesis it is much more important to be able to specify the **absolute configuration** at a given stereogenic centre. This is done using the *Cahn–Ingold–Prelog* system[1] first introduced in 1951, the same year in which the absolute configuration of a chiral compound was first determined. It is assumed that the reader will already be familiar with the basic features of this system and so only a brief summary is given here together with some special applications.

The procedure for specifying the absolute configuration at a tetrahedral centre first involves placing the four groups in order of priority. The priority is based on atomic number or, for isotopes, atomic weight, and any two groups are compared by working out from the tetrahedral centre until a difference between them is found. Multiple bonds are treated by introducing duplicate atoms as exemplified by the carboxylic acid and cyano groups shown.

$$-\text{C}\begin{matrix}\diagup\text{O}\\\diagdown\text{OH}\end{matrix} \quad = \quad -\underset{\underset{\text{OH}}{|}}{\overset{\overset{(\text{O})\ (\text{C})}{|\ \ \ |}}{\text{C}}}-\text{O} \qquad -\text{C}\equiv\text{N} \quad = \quad -\underset{\underset{(\text{N})\ (\text{C})}{|\ \ \ |}}{\overset{\overset{(\text{N})\ (\text{C})}{|\ \ \ |}}{\text{C}}}-\text{N}$$

THE DESCRIPTION OF STEREOCHEMISTRY

This is only done to bring each atom up to four ligands including any lone pairs which may be present and is **not** necessary for S=O or P=O double bonds, for example, which are treated as single bonds. Two additional points to note are that the atomic weight priority readily accounts for isotopes (e.g. D > H) and a lone pair of electrons is considered as an atom of atomic weight zero (i.e. lowest priority). Once the groups have been placed in order of priority (1–4) the tetrahedral centre is viewed with the lowest priority group (4) at the back and the arrangement of the remaining three groups considered. If the groups lie in order of decreasing priority (1–2–3) in a clockwise arrangement the configuration is denoted R and if the arrangement is anticlockwise the configuration is S. The application of this system to some of the chiral molecules considered in chapter 1 is shown below.

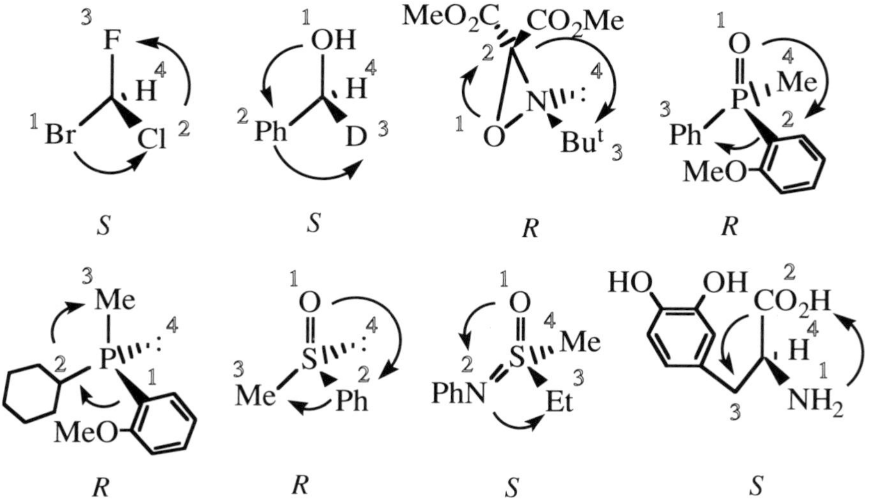

It is worth noting that most of the naturally occurring sugars have the absolute configuration (**1**) at the centre shown and are therefore all R at this centre while the naturally occurring amino acids all have the configuration (**2**) which is S [except for cysteine (R = CH$_2$SH) where the presence of sulphur causes R to take priority over CO$_2$H making this the R enantiomer].

The Fischer nomenclature for these two series of compounds, D for (**1**) and L for (**2**), which was based on their chemical correlation to the two enantiomers of glyceraldehyde, has been largely superseded by the Cahn–Ingold–Prelog system.

2.2 Axially chiral compounds

The absolute configuration of compounds with a stereogenic axis, such as the allenes and hindered biaryls, can also be specified as *R* or *S* using the Cahn–Ingold–Prelog system[2] but an additional rule is needed because, as we saw in section 1.9, two of the groups may be the same. The groups attached to the stereogenic axis are first placed in order of priority at each end of the molecule separately. The groups on the end of an allene or the *ortho* positions of a biaryl actually take up a distorted tetrahedral arrangement in space. To assign the configuration we view this tetrahedron with the lower priority group on one of the ends at the back (it turns out not to matter which end is chosen) and then consider the arrangement of the remaining groups

THE DESCRIPTION OF STEREOCHEMISTRY 25

in the order: high priority (front) > low priority (front) > high priority (back). As before if these appear clockwise the configuration is *R* and if they are anticlockwise it is *S*. The application of this rule is illustrated by the examples shown on p. 24.

2.3 Compounds with more than one stereogenic unit

For a molecule with more than one stereogenic unit the absolute configuration can be specified by giving the configuration of each unit using the Cahn–Ingold–Prelog system. Thus for example (–)-threonine (**3**) is (2*S*, 3*R*)-2-amino-3-hydroxybutanoic acid and compound (**4**) has the configuration (2*S*, 3*S*).

In considering a diastereoselective reaction, however, we are frequently concerned with the **relative configuration** of two stereogenic units rather than their absolute configuration. This was previously denoted by the prefixes e*rythro-* and *threo-* but these were only applicable to certain types of compound and their usefulness was further diminished when a redefinition was recently proposed[3] (and adopted by some but not all authors) which amounted to a reversal of their meaning. There are now two systematic methods of specifying relative configuration, both of which rely on the Cahn–Ingold–Prelog system.

The first of these, introduced by Seebach and Prelog in 1982,[4] involves comparison of the configuration of the stereogenic units present, once they have been assigned as *R/S*. The combinations (*R,R*) and (*S, S*) are denoted by *l* (like) and (*R, S*) and (*S, R*) by *u* (unlike). Thus, for example, compound (**3**) which has the configuration (2*S*, 3*R*) is a *u* diastereomer and (**4**), with the (*S, S*) configuration, is *l*.

It is important to remember that *l* and *u* only specify the *relative* configuration and both (–)-threonine (**3**), which is (2*S*, 3*R*), and its enantiomer (+)-threonine, which is (2R, 3S), are *u* isomers. For compounds with more than two stereogenic units, these can be compared sequentially following the normal numbering of the compound. Thus a compound with absolute configuration (1R, 2S, 4S, 6R) (and its enantiomer) would be described as *u*, *l*, *u*. With several stereogenic units this becomes increasingly open to ambiguity however, and in these cases the relative configuration is probably better specified using the second system described below.

As already mentioned, any term which only describes the relative configuration of the stereogenic units present actually applies to a pair of enantiomers and this forms the basis of a highly systematic way to specify relative configuration. This consists quite simply of choosing the enantiomer which has the *R* configuration in the first position and adding asterisks to its absolute configuration. Thus, for example, the *u* diastereomer of threonine, which includes the (2S, 3R) and (2R, 3S) enantiomers, can be denoted (2R*, 3S*) and compound (**4**) with absolute configuration (2S, 3S) (as well as its enantiomer) can be denoted (2R*, 3R*). Although this system may seem unwieldy it can readily be extended to cover any number of stereogenic units without ambiguity and it is the method used by *Chemical Abstracts*.

Finally in this section it is worth considering one further example, that of tartaric acid, which illustrates several important points. Since the compound has two stereogenic centres we might expect four possible stereoisomers (**5**–**8**). In fact we can see that (**7**) and (**8**) are identical since they differ only by a rotation of 180° about a horizontal

(5)   (6)   (7)   (8)

axis and can be readily superimposed. Furthermore, the single form of tartaric acid represented by (**7**) and (**8**) is *achiral* since these molecules can be superimposed on their mirror images. Compounds of this type which are achiral and have a plane of symmetry present are called ***meso* compounds**. Thus we have three forms of tartaric acid: (+), which happens to correspond to (**5**) (see section 1.13), (−) which corresponds to (**6**) and *meso* which corresponds to (**7**) or (**8**). If we now consider the absolute configurations of these isomers, the

THE DESCRIPTION OF STEREOCHEMISTRY

Cahn–Ingold–Prelog system gives the description (R, R) for (**5**), (S, S) for (**6**) and (R, S) for both (**7**) and (**8**). Note that no numbers are needed here since the two positions are equivalent. Thus both (**5**) and (**6**) are *l* or (R*, R*) and the *meso* form is *u* or (R*, S*). This example highlights the fact that chirality is a property of the whole molecule and we can only determine whether or not a molecule is chiral by testing whether it can be superimposed on its mirror image. Thus, although *meso*-tartaric acid contains two stereogenic centres, whose absolute and relative configurations can be perfectly well described using the Cahn–Ingold–Prelog system, it is not chiral. Likewise the fact that each tetrahedral centre is joined to four different groups does not make the molecule chiral.

Even though they are achiral, *meso* compounds are of some value as substrates in asymmetric synthesis, since, upon enantioselective reaction, they give chiral products with two contiguous stereogenic centres (see sections 6.5.2 and 6.5.3).

2.4 The topicity of enantioselective reactions

In considering an enantioselective reaction it is useful to be able to specify not only the absolute configuration of the final product but also the direction in which the reaction occurs, its **topicity**. This is done by applying the Cahn–Ingold–Prelog system to the enantiotopic faces or groups of the reactant molecule. For a trigonal planar reactant, the face from which the three groups appear clockwise in order of decreasing priority (assigned as before) is described as the *Re* face and the opposite face from which they appear anticlockwise is described as *Si*. Taking the example of the reduction of acetophenone again, the front face as drawn in (**9**) is *Si* and the back face is *Re*. The topicity of the asymmetric reduction can now be described according to which face hydride attacks from. Thus *Re* reaction with H⁻ gives alcohol (**10**) while *Si* reaction gives the enantiomer (**11**).

It is important to note that there is no direct connection between *R/S* and *Re/Si* since this will depend on the priority of the attacking group relative to those already present. The topicity of the reaction and the absolute configuration of the product must be assigned separately. This point is illustrated by the fact that *Re* reaction of (**9**) with H⁻ gives the *S* product (**10**) but *Re* reaction of (**9**) with EtMgBr would give (**12**) which has the *R* configuration.

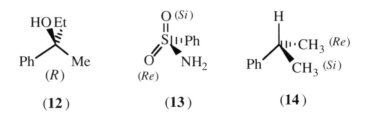

The *Re/Si* system can also be used to specify enantiotopic groups or atoms. In this case the group from which the other three appear clockwise in order of decreasing priority is described as *Re* while the other, from which they appear anticlockwise is *Si*. The two examples (**13**) and (**14**) illustrate the application of this rule.

2.5 The relative topicity of diastereoselective reactions

The *l/u* system of nomenclature, introduced by Seebach and Prelog to describe the relative configuration of diastereomers, can be extended to specify the **relative topicity** of diastereoselective reactions.[4] This involves comparing the *R/S* and *Re/Si* specifications of the stereogenic units and/or enantiotopic faces or groups involved. The combinations (*R, Re*), (*S, Si*), (*Re, Re*) and (*Si, Si*) are denoted *lk* (like) while (*R, Si*), (*S, Re*), (*Re, Si*) and (*Si, Re*) are denoted *ul* (unlike). To illustrate the use of this system we will again take the examples of section 1.11 which are representative of the two major types of diastereoselective reaction.

In the first case we begin by assigning the diastereotopic faces of the carbonyl group in (**15**) as *Re* and *Si* according to the procedure described in section 2.4 and neglecting for the moment the presence of the stereogenic centre. This initially makes the front face as shown in (**15**) *Re* and the back face *Si*. If we now combine these assignments with the fact that the stereogenic centre is *S*, the final descriptions are *ul* for the front face (*Re,S*) and *lk* for the back face (*Si,S*). Thus reduction by *ul* attack of H⁻ gives product (**16**) which has the *l* relative configuration while *lk* attack of H⁻ gives the *u* product (**17**). Once again it is clear that the relative topicity of the reaction (*lk/ul*) and the relative configuration of the product (*l, u*) are not directly related.

(**17**)   (**15**)   (**16**)

For reactions in which two stereogenic units are formed simultaneously we first assign the *Re* and *Si* faces of each prochiral reactant molecule. For this purpose it is important to choose the actual reacting species rather than a precursor. For the reaction of benzaldehyde with the anion of ethyl propionate this gives the results shown in (**18**) and (**19**).

(**18**)   (**19**)

The relative topicity of any of the four possible modes of reaction can now be specified by comparing the *Re/Si* designations of the reacting faces and assigning the reaction as *lk* or *ul* as before. Thus, for example, reaction between the *Si* face of (**18**) and the *Re* face of (**19**) as shown in (**20**) is denoted *ul* and gives rise to the major product (**21**).

2.6 Further points on correct terminology

In this section we draw attention to several common errors in terminology relating to asymmetric synthesis.

The first point, which should be obvious to most readers, is that a tetrahedral centre is properly represented by (**22**) in which the

two out of plane bonds are in adjacent quadrants of the figure. The figure (**23**) represents a *square planar* arrangement of bonds around the centre and is not acceptable as a representation of a tetrahedron. The figures (**24**) and (**25**) are poor representations of the tetrahedral shape and actually correspond to opposite enantiomers. The arrangement of groups around a tetrahedral centre should always be as shown in (**22**).

A second area concerns the use of the word *chiral*. This has a specific definition and can only be properly applied to three-dimensional objects. Thus it is incorrect to refer to a 'chiral centre/axis' or a 'chiral synthesis'. The former can be a 'chirality centre/axis' or a 'centre/axis of chirality' although for most purposes we prefer to replace these terms by 'stereogenic centre/axis' (which is, however, *not* equivalent).[5] As a collective term for enantio- and diastereoselective synthesis we have retained the term 'asymmetric synthesis' due to its common usage and the less specific meaning of asymmetric.

It is important to realise that chirality is a symmetry property of a whole molecule and cannot be localised in a particular centre or group (although it may be associated with the presence of a particular stereogenic unit). For this reason the *intramolecular* **transfer of chirality** is an impossibility. Transfer of chirality can only occur between two molecules and is quite uncommon. [An example is the enantioselective Meerwein–Ponndorf–Verley reduction of a ketone with a chiral alcohol.] What some authors refer to incorrectly as transfer of chirality is, in fact, **retention** of chirality in the course of modification of the stereogenic units present.

References and notes

1. R.S. Cahn, C.K. Ingold, and V. Prelog, *Angew. Chem., Int. Ed. Engl.,* 1982, **5**, 385; V. Prelog and G. Helmchen, *ibid.,* 1982, **21**, 567.
2. The absolute configuration of axially chiral compounds can alternatively be specified by M and P which also apply to planar chiral compounds: see reference [1].
3. For a discussion of the confused situation here see C.H. Heathcock in *Asymmetric Synthesis*, Vol. 3, J.D. Morrison, ed., Academic Press, Orlando, 1984, p. 112.
4. D. Seebach and V. Prelog, *Angew. Chem., Int. Ed. Engl.,* 1982, **21**, 654.
5. K. Mislow and J. Siegel, *J. Am. Chem. Soc.,* 1984, **106**, 3319.

The table below gives a summary of the approved descriptive terms introduced in chapters 1 and 2:

For description of:	Approved terms	Older or disfavoured terms
Enantiomers	(+)-, (−)-	*d/l*
Enantioselectivity	% e.e.	
Diastereoselectivity	% d.e. or diastereomer ratio	% d.s.
Absolute configuration	*R/S*	D/L
Relative configuration	*l/u* or *R*, R*/R*, S**, etc.	*erythro/threo*
Enantiotopic faces or groups	*Re/Si*	proR/proS
Topicity of enantioselective reactions	*Re/Si*	
Relative topicity of diastereoselective reactions	*lk/ul*	

3 Analytical methods: determination of enantiomeric purity

D. PARKER and R.J. TAYLOR

The development of accurate non-polarimetric methods for the determination of enantiomeric purity, which began in the 1960s, has been critical in the development of asymmetric synthesis, allowing precise and reliable assessment of the degree of selectivity achieved in a given reaction. In this chapter the main methods, together with some of their advantages and disadvantages, are described.

3.1 Polarimetric methods

The classical method of determining the enantiomeric purity of a sample is to measure its optical purity using a polarimeter. The optical purity is derived by expressing the measured optical rotation $[\alpha]$ as a percentage of the optical rotation of the pure enantiomer. Measurements are taken under standardised conditions which indicate the wavelength of the incident plane-polarised light, the temperature of measurement, the solvent and the concentration of the solute in grams per 100 cm^3. For example, enantiomerically pure (R)-mandelic acid has $[\alpha]_D = -155.4°$ ($c = 2.913$, 95% EtOH), where D refers to the wavelength of the sodium-D line emission (589 nm). The measurement is quick and straightforward and requires a relatively inexpensive polarimeter. The sample to be assayed is dissolved (or mixed) in an achiral solvent to give a solution of known concentration. This is placed in a cell of known path length within the polarimeter (the path length is often 10 cm) and plane-polarised light from a sodium vapour lamp is allowed to pass through the solution. The interactions of the constituent left- and right-hand circularly polarised vectors (of the resultant plane-polarised light) with the chiral medium are non-equivalent. This results in the plane of the plane-polarised light being rotated, the extent and sense of which are measured directly. The specific rotation is defined to be:

$$[\alpha]_\lambda^t = \frac{100 \cdot \alpha'}{l \cdot c}$$

α'	=	observed rotation
l	=	cell path length in dm (usually $l = 1$)
c	=	concentration in g per 100 cm^3 of solvent
t	=	temperature (Celcius)
λ	=	wavelength of incident light (nm)

Although this method remains the most commonly used technique for assaying enantiomeric purity, it does possess some disadvantages. The sample under analysis must be homogeneous, devoid of trace chiral impurities and should be isolated from a reaction mixture without accidental enantiomeric enrichment. Particular care should be taken **not** to crystallise a solid sample, for example, as enantiomers often crystallise at different rates in a chiral medium (such as in the presence of the other enantiomer). However, distillation or chromatographic methods of purification are acceptable. For compounds with low optical rotatory power, large samples may be required to produce measurable optical rotations. As mentioned already in chapter 1, such a case arises for compounds which are chiral by virtue of isotopic substitution. It is self-evident that the maximum rotation of the pure enantiomer must be known with certainty. There are many cases in the scientific literature where absolute optical rotations have been reported which are incorrect, or are interpreted incorrectly. For example, prior to 1974, the rotation of optically pure (+)-3-methylcyclopentene was believed to be $[\alpha]_D = +78°$. Following independent measurement of the enantiomeric purity using a chiral gas chromatographic method, the correct rotation was proved to be $[\alpha]_D = +174.5°$. More recently certain workers assumed that the optical rotation of *exo*-2-norbornane carboxylic acid was $[\alpha]_D = -10.7°$ ($c = 1.0$, EtOH). Unfortunately they had misread the original article and an independent NMR method was required to confirm that the 'hidden' value, $[\alpha]_D = -27.8°$, was indeed correct. It is perhaps worth reflecting that at the end of the last century, Berthelot was convinced that styrene was optically active.

As mentioned earlier, optical rotations are particularly sensitive to temperature and concentration, and it has been estimated that errors in measured rotations from these combined effects are at least ±4%.

The temperature variation is caused by the change in the volume of the sample liquid or solution resulting in a direct change in the number of chiral molecules in the optical path. The interaction of solute molecules with each other and with solvent molecules is also sensitive to temperature, altering the relative populations of stereochemically important conformations. The specific optical rotation is not always linearly proportional to concentration. Interactions between solute molecules may lead to non-linear rotations in concentrated solution. It is *essential* therefore to compare the same concentration of solution (in the same solvent) when comparing rotations on an absolute basis, and to report the concentration of solution when recording optical purity data for new compounds. Preferably at least two determinations should be made at different concentrations to indicate whether or not the dependence of optical rotation on concentration is linear. An extreme example of such concentration dependence is afforded by the behaviour of (**1**) in chloroform solution.[1] At a concentration of 6.3 g/100 ml, the *S* enantiomer of (**1**) has a rotation of zero degrees, at concentrations above 6.3 the observed rotation is negative, but below 6.3 the rotation is positive. Evidently selective molecular association of the polar diacid occurs in the non-polar solvent.

$$\text{Et} \underset{\text{CH}_2\text{CO}_2\text{H}}{\overset{\overset{\displaystyle \text{CO}_2\text{H}}{|\text{Me}}}{\diagup\!\!\!\diagdown}}$$

(**1**)

Although optical and enantiomeric purities are usually equated, this is not necessarily correct. Indeed it was with (**1**) that the *inequivalence* of enantiomeric purity and optical purity was first demonstrated unequivocally. If optical rotation does not vary linearly with concentration (and this may occur even in polar solvents) then an alternative method for measuring enantiomeric excess must be sought.

It is often possible to correlate absolute configuration with the sense of optical rotation, and several empirical rules have been devised over the years based on this premise. Unfortunately this is not always the case. There are several instances where a closely related reactant and product with the same absolute configuration rotate the plane of the plane-polarised light in different directions.

In summary, measurements of optical rotation may be used to determine enantiomeric purity, but only if the readings are taken carefully with a homogeneous sample under controlled conditions and with deference to the factors which affect measured rotations (temperature, concentration, wavelength, solvent). The determination of the enantiomeric purity on the basis of optical rotation measurements should be confirmed by an independent method before assigning absolute rotations.

Mention must also be made of the more powerful polarimetric techniques of optical rotatory dispersion (ORD) and circular dichroism (CD). As alluded to above, the optical rotation, α, varies as a function of the wavelength of the light. Unless the molecule has a chromophore, the ORD curve (plot of α with wavelength) declines monotonically with increasing wavelength. In the vicinity of an electronic absorption, however, the ORD curve can undergo a dramatic inversion of sign known as the Cotton effect. In favourable cases it is possible to deduce the *absolute* configuration of the asymmetric molecule from the sign and magnitude of the Cotton effect. CD is a complementary technique in which the differential absorption of polarised light by the chromophore is studied.[2]

3.2 Gas chromatographic methods

An attractive method for the analysis of mixtures of enantiomers is chiral gas chromatography (GC). This sensitive method is unaffected by trace impurities, and is quick and simple to carry out. The premise upon which the method is based is that molecular association may lead to sufficient chiral recognition that enantiomer resolution results. The method uses a chiral stationary phase which contains an auxiliary resolving agent of high enantiomeric purity. The enantiomers to be analysed undergo rapid and reversible diastereomeric interactions with the stationary phase and hence may be eluted at different rates. There are certain limitations to the method, some of which are peculiar to the gas chromatographic method. The sample should be sufficiently volatile and thermally stable, and, of course, should be quantitatively resolved on the chiral GC phase. Occasionally this means that enantiomeric mixtures need to be derivatised (with an achiral reagent) prior to GC analysis (e.g. trifluoro-acetylation of amino acids). In this case the lack of racemisation during derivatisation must be carefully proven. As the method involves enantiomers rather than diastereomers, the enantiomeric purity of the sample is insensitive to chemical, physical or analytical manipulation prior to analysis. The observed enantiomeric ratio is independent of the enantiomeric purity of the chiral support, although a lower column enantiomeric purity results in poorer resolution, i.e. the separation factor, α, is reduced.

Once a quantitative resolution of enantiomers on the chiral stationary phase has been achieved, inspection of the chromatogram gives useful information directly. The peak ratio gives a precise and quantitative measure of the enantiomeric composition of the sample. Such measurements may be effected with a high degree of accuracy ($\pm 0.05\%$) so that precise data may be obtained. This is particularly useful for analysing mixtures which are either highly enriched, $95\% <$ e.e $< 100\%$, or close to the racemic limit. The resolution of enantiomers by chromatography relies upon the disparity between the free energies of formation of the transient diastereomeric intermediates formed during elution. The assignment of absolute configuration will therefore involve the correlation of molecular configuration with the order of enantiomer elution. Although such correlations often work in closely structurally related series, exceptions that demonstrate the limitations of the approach have been defined. The other important GC parameter is the peak separation factor, α. This provides a thermodynamic measure of the degree of chiral recognition between the racemic solute and the chiral support. As little as 0.1% enantiomeric excess is detectable by chiral GC, corresponding to an energy difference of $\Delta\Delta G^{\#}_{298} = 16.7$ J mol^{-1} between the two diastereomeric intermediates.

Most of the reported applications of this technique involve the direct resolution of *derivatised* enantiomers on chiral stationary phases. The earliest successful resolution of *N*-trifluoroacetyl amino acid esters on glass capillary columns coated with *N*-trifluoroacetyl-*S*-isoleucine lauryl ester (**2**) was effected in 1966. Superior stationary phases developed quickly, such as (**3**), in which the additional amide bond may interact *via* hydrogen bonding. These chiral supports work well for the separation of *N*-perfluoracylated derivatives of a given amino acid. However, their low thermal stability (190°C maximum) and appreciable volatility have inhibited their use for the simultaneous

38 ASYMMETRIC SYNTHESIS

$$\underset{(\mathbf{2})}{F_3C-\underset{\parallel}{C}-\underset{H}{N}-\overset{H}{\underset{\vdots}{C}}(Bu^s)-CO_2C_{12}H_{25}{}^n}$$

$$\underset{(\mathbf{3})}{H_{23}C_{11}{}^nO-\underset{\parallel}{C}-\underset{H}{N}-\overset{H}{\underset{\vdots}{C}}(Pr^i)-CO_2Bu^t}$$
$$(H_{43}C_{21}{}^nO)$$

$$\underset{(\mathbf{4})}{\left[O-\underset{\underset{Me}{|}}{\overset{\underset{Me}{|}}{Si}}\right]_7 O-\underset{\underset{H_2C-CH(Me)-C(=O)-NH-CH(Pr^i)-CO_2Bu^t}{|}}{\overset{\underset{Me}{|}}{Si}}-}$$

resolution of a series of naturally occurring amino acids in one run, when temperature programming is required. With this in mind the fluid polymeric phase (**4**) (Chirasil-Val) has been developed by coupling (**3**) to a copolymer of methyl siloxane and (2-carboxypropyl)-methyl siloxane. The simultaneous GC separation of the enantiomers of all protein amino acids (as their N-pentafluoropropanoyl isopropyl esters) using (**4**) coated on a glass capillary column. Figure 3.1 provides an impressive example of the power of this technique. Many

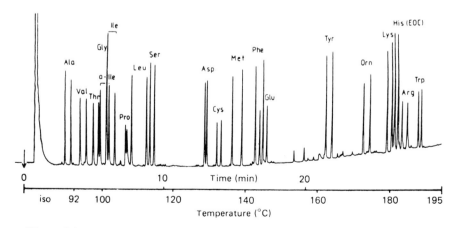

Figure 3.1 Gas chromatographic resolution of enantiomers of racemic protein amino acids as their N-pentafluoropropanoyl isopropyl estars on (**4**). (Boyer, E.; Frank, H. In *Modification of Polymers*; Carraher, Jr., C. E.; ACS Symposium Series 121; American Chemical Society: Washington, DC, 1980; p. 66.)

other diamide sub-units have been coupled to polysiloxane backbones leading, for example, to resolution of trifluoroacetyl derivatives of secondary amines and the resolution of alcohols and α-hydroxy acids as their isopropylurethane derivatives after derivatisation with isopropyl isocyanate. Such a method established that commercial (S)-ethyl lactate contained 1.67% of the R enantiomer, i.e. it had an e.e. of 96.66%.

The direct resolution of *underivatised* enantiomers has been effected with the aid of some metal-containing chiral stationary phases. Particular attention has been paid to the resolution of chiral alkenes and epoxides using stationary phases derived from chiral metal chelate complexes. The rhodium dicarbonyl β-diketonate complex (**5**), when dissolved in squalene and coated onto a capillary column, permitted the quantitative enantiomer resolution of 3-methylcyclopentene while methyl oxirane was similarly resolved

(**5**)

(**6**) R = CF_3
(**7**) R = $CF_2CF_2CF_3$

using the nickel camphorato complex (**6**). Using the related nickel complex (**7**), the direct enantiomer resolution of volatile chiral alcohols and ketones (such as menthol and isomenthone) has been achieved using methylsilicone as a solvent in the preparation of the capillary column. The development of this direct technique is in its early stages and it is anticipated that designed chiral metal chelates will broaden the scope of this simple method.[3]

3.3 Liquid chromatographic methods

The development of rapid, simple liquid chromatography methods for the assay of enantiomeric purity has perhaps been the most important development in the analysis of chiral compounds in the last ten years. Using analytical high performance liquid chromatography (HPLC), the enantiomeric purity of thousands of compounds has been assessed. The separability factor, α*, for two components in a HPLC chromatogram depends upon the band shape and is related directly to the efficiency of the column, i.e. flow rate, particle size, sample size and quality of packing. Efficient HPLC systems produce good separations for two components having $\alpha \geq 1.05$. As in chiral GC, enantiomer separation requires a chiral agent. This may involve pre-column derivatisation (in the absence of racemisation) with a chiral derivatising agent to give chromatographically separable diastereomers. A full description of the chromatographic separation of diastereomeric species formed by this method is given elsewhere in detail,[4] but is beyond the scope of this chapter. A more direct and preferred method is to induce short-term diastereomeric interactions of the two enantiomers with a chiral stationary phase. The diastereoisomeric complexes formed will have non-identical stabilities and hence elute at different times. An alternative is to use an achiral support and to elute with a chiral eluant.

The direct method is more elegant and less problematic than the indirect approach. The pre-column derivatisation of enantiomeric mixtures is not an absolute method and less than 100% enantiomeric purity of the derivatising agent or kinetic resolution may adversely affect the measure of enantiomeric purity obtained. In contrast, the direct methods are absolute in that no external standard of enantiomeric purity is required. Like many of the NMR methods, direct chromatographic analysis gives a weighted time average of dynamic processes. A less than enantiomerically pure chiral stationary phase or eluant will affect the separation but not the relative size of the eluting bands arising from the solute enantiomers. Any reduction in the enantiomeric purity of the chiral eluant or stationary phase reduces α by decreasing K_2 and increasing K_1 until at the racemic limit $K_1 = K_2$ and $\alpha = 1$. Until very recently the advantages of the direct approach had been tempered by the lack of commercially available chiral stationary phases. This situation has now changed and chiral stationary phases based upon 9-anthryltrifluoroethanol and N-(2-naphthyl)-α amino acids in particular should find general use in the near future. Both require solute

---

* The volume of solvent required to elute a non-retained sample is equal to the void volume of the column. In order to elute a retained sample, an additional volume of solvent is required. The ratio of this additional volume to the void volume is the capacity ratio K, and the separability factor, α, for two components is K_2/K_1. Alternatively α is the ratio of retention times measured from the time of elution of a non-retained solute.

derivatisation to enhance detectability or improve α, and this often involves O, N or S acylation, e.g. with 3,5-dinitrobenzoyl chloride or 3,5-dinitrophenyl isocyanate. As in chiral GC methods, absolute configuration can be determined if a sample of known configuration is available for reference. Many absolute configurations can be assigned with some confidence based upon elution order, provided that the molecular configuration of the chiral agent and mechanism of the chromatographic diastereomer separation are known.

Before synthetic chiral stationary phases were developed, attempts were made to use naturally occurring chiral materials for the stationary phase. Quartz, wool, lactose and starch were inadequate but triacetylated cellulose has met with some success. The synthetic stationary phases introduced by Pirkle are able to interact with solute enantiomers in three ways, one of which is stereochemically dependent. Typically these interactions are based on hydrogen bonding, charge transfer (π-donor–π-acceptor based) and steric repulsive types. An independent chiral stationary phase therefore consists of chiral molecules each with three sites of interaction bound to a silica (or other) support. Early work in this area demonstrated that S-arginine bound to Sephadex would resolve 3,4-dihydroxy-phenylalanine, and that direct resolution of chiral helicenes could be accomplished with columns packed with 2-(2,4,5,7-tetranitro-9-fluorenylideneaminoxy)-propionamide or tri-β-naphthol-diphosphate amide. Amino acid esters have also been resolved with a silica bound chiral binaphthyl crown ether, but better separations are achieved with N-acylated amino acid derivatives with amino-acid derived chiral stationary phases.

Two of the most useful chiral stationary phases are (**8**) and (**9**). Mercaptopropyl silanised silica was alkylated to prepare (**8**) and this stationary phase permits the resolution of all alkyl 2,4-dinitrophenyl sulphoxides and many 3,5-dinitrobenzoyl derivatives of chiral alcohols and amines, provided that these compounds possess an additional basic site to interact with the acidic methine proton of the trifluoroethanol moiety. The diastereomeric interactions that allow a column derived from chiral A to resolve racemic B, also allow a column derived from chiral B to resolve racemic A. This reciprocality has been used in the design of new chiral stationary phases and such an approach led to the development of (**9**) derived from (*S*)(–)-*N*-(2-naphthyl)valine. Dinitrobenzoyl or dinitrophenyl carbamoyl derivatives of a broad spectrum of chiral amines, amino alcohols, alcohols and thiols may be assayed with this chiral stationary phase. It should prove to be of considerable use to those engaged in asymmetric synthesis or for those concerned with monitoring the enantiomeric composition of drugs or metabolites in body fluids.

The separation of enantiomers on an achiral support using a chiral eluant has been restricted to chiral amino acids. A metal ion (usually Cu^{2+} or Ni^{2+}) is also present in the solution and forms diastereomeric complexes containing the chiral eluant (e.g. (*S*)-proline) and either of the substrate enantiomers. The method is therefore a variant of ligand exchange chromatography, and an example of its use in the separation (and hence assay) of amino acids is given in Figure 3.2. When enantiomers are separated on an achiral column using a chiral eluant, the mechanism of separation is complex. It may result from a combination of the different stability of the diastereomeric complexes in the mobile phase, differential adsorption of the isomeric complexes or adsorption of the chiral eluant onto the achiral support so that the support effectively acts as a chiral stationary phase. Thus correlation of absolute configuration with elution order may be inappropriate if extrapolated too far from defined examples.

3.4 NMR spectroscopy

Enantiomers cannot be distinguished in an achiral medium by their NMR spectra because their resonances are chemical shift equivalent (isochronous). In contrast, diastereomers may be distinguished because certain resonances are chemical shift non-equivalent (anisochronous). Determination of enantiomeric purity using NMR requires the intervention of a chiral auxiliary to convert an enantiomeric mixture into a mixture of diastereomers. Provided that the magnitude of the observed chemical shift non-equivalence is sufficient to give baseline

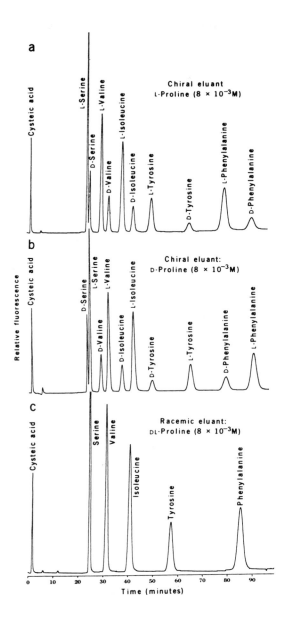

Figure 3.2 The separation of (*R*)- and (*S*)-amino acid enantiomers by ligand exchange chromatography using proline and copper (II) sulphate solution as the chiral eluant.

resolution, integration of the appropriate signals gives a measure of the diastereomeric composition. This can be directly related to the enantiomeric composition of the original mixture.

Three types of chiral auxiliary are widely used. Chiral derivatising agents (CDAs)[5] form diastereomers while chiral solvating agents (CSAs)[6] and chiral lanthanide shift reagents (CLSRs)[7] form diastereomeric complexes *in situ* with the substrate enantiomers.

An effective chiral auxiliary should induce significant NMR chemical shift anisochronicity in as large a range of substrates as possible. Further, if the sense of non-equivalence is consistent in a series of compounds, then once a standard of known stereochemistry has been studied the absolute configuration of the major and minor enantiomers present in chemically similar unknown mixtures can be deduced from the NMR spectra.

The magnitude of the chemical shift non-equivalence is proportional to the size of the applied magnetic field. Lowering the temperature at which the spectrum is recorded can accentuate the anisochronicity between diastereomers. The use of non-polar solvents such as d-chloroform and, in particular, aromatic solvents such as d_6-benzene or d_8-toluene offers considerable advantages. This effectively excludes the application of NMR methods for the assay of the enantiomeric purity of substrates which are only soluble in polar solvents like d_6-DMSO. It is unfortunate that numerous pharmacologically important compounds fall into this category. In such cases chiral GC or chiral HPLC methods may afford viable alternatives. Proton, ^{13}C, ^{19}F and ^{31}P are the most frequently studied nuclei. It is important to note that measured integrals will only report reliably on the enantiomeric purity in fully relaxed spectra free from any saturation effects.

3.4.1 Chiral derivatising agents (CDAs)

An enantiomeric mixture can be converted to a pair of diastereomers prior to NMR analysis by reaction with a chiral derivatising agent. Unlike chiral solvating agents and chiral lanthanide shifts reagents which form diastereomeric complexes *via* reversible equilibria, a CDA forms discrete diastereomers free from the effects of chemical exchange. As a result the magnitude of the chemical shift non-equivalence, $\Delta\delta$, is typically five times larger than that observed in the presence of a chiral solvating agent (CSA).

ANALYTICAL METHODS 45

There are a number of disadvantages in the use of a CDA.

(a) An additional chemical reaction is necessary before NMR analysis of the sample can be performed.
(b) Kinetic resolution due to differential reaction rates of the substrate enantiomers with the CDA can lead to spurious enantiomeric purity results.
(c) The stereochemical integrity of the derivatisation and any purification steps thereafter must be rigorously established. There must be no possibility of racemisation or accidental enrichment accompanying these processes.
(d) The CDA must be enantiomerically pure; the presence of a small quantity of the opposite enantiomer of the CDA will reduce the observed enantiomeric excess value.

Despite these problems and limitations, chiral derivatisation remains the most widely used NMR technique for enantiomer resolution. CDAs are usually simple, multifunctional compounds which are often commercially available and claimed to be pure enantiomers. Derivatisation frequently involves esterification or amidation under non-racemising conditions. The method is reliable, which is not always the case with CSAs or CLSRs. In addition, the sense of non-equivalence is consistent allowing assignment of absolute configuration on the basis of chemical shift to be made with some confidence. Diastereomeric anisochronicity is usually sufficient to permit enantiomeric excess measurement to within ±1% even with small applied magnetic fields (≤ 100 MHz).

Chiral acids react with chiral alcohols or amines to form diastereoisomeric esters or amides respectively. Mislow and Raban[8] first described chemical shift non-equivalence in the proton NMR spectra for diastereoisomeric 1-methylphenylethanoic acid esters of 1-(2-fluorophenyl)-ethanol and observed some racemisation during the reaction. A systematic study was made thereafter of a series of substituted phenylethanoic acids as CDAs for the assay of alcohols (Table 3.1). Epimerisation α to the acid carbonyl was the cause of the racemisation observed by Mislow and Raban. In order to avoid this problem Mosher developed α-methoxy-α-trifluoromethyl-phenyl-acetic acid (MTPA), (10), as a CDA.[9] It is stable to racemisation because it lacks an α-hydrogen. Induced ^{1}H chemical shift non-equivalence is typically 0.15 ppm (in CDCl$_3$, 298 K). A ^{19}F NMR study

Table 3.1 Proton chemical shift non-equivalence for ester derivatives of chiral alcohols.

Alcohol used		Mandelate		O-Methylmandelate		MTPA ester		
		\multicolumn{7}{c	}{magnitude of Δδ in ppm (sense of non-equivalence)[a]}
R^1	R^2	R^1	R^2	R^1	R^2	R^1	R^2	CF_3[b]
Et	Me	0.30(−)	0.18(+)	0.08(−)	0.15(+)	0.10(+)	0.13(−)	0.08(+)
n-Hex	Me	*	0.13(+)	*	0.05(+)	0.08(+)	0.08(−)	0.32(+)
Pr^i	Me	0.15(−)	0.20(+)	0.12(−)	0.12(+)	0.08(+)	0.08(−)	0.17(+)
Bu^t	Me	0.15(−)	0.24(+)	0.10(−)	0.10(+)	0.07(+)	0.07(−)	0.22(+)
Bu^t	Et	0.22(−)	0.34(+)	0.13(−)	0.10(+)			
Bu^t	Bu^n	0.32(−)	*			0.06(+)	*	0.25(−)
CF_3	Me	0.17(−)	0.21(+)	*	0.12(+)	0.12(+)	0.41(−)	0.28(+)
Ph	CF_3	*	0.06(+)			*	0.33(−)	0.57(+)
Ph	Me	*	0.15(+)	*	0.08(+)	*	0.06(−)	0.51(+)
Ph	Et	*	0.28(+)	*	0.12(+)	*	0.08(−)	
Ph	Pr^i	*	0.25(+)	*	0.08(+)			0.35(+)
Ph	Bu^t	*	0.22(+)	*	0.08(+)	*	0.05(−)	0.50(+)

[a] (+) = high frequency, (−) = low frequency [b] ^{19}F signal
* = no splitting observed

is also possible and this has the advantage that the ^{19}F spectrum contains only two peaks: one for each enantiomer. Other acids, in particular, camphanic acid (**11**), and *O*-acetyl-mandelic acid, (**12**), are also effective CDAs for alcohols and α-deuteriated amines. α-Amino-acids and α-hydroxyacids can also be assayed using MTPA in association with a silver shift reagent Ag(fod).

In the reciprocal experiment, the chiral alcohol methylmandelate, (**13**), is used as a CDA for the study of the enantiomeric purity of acids. Esterification is effected without racemisation with *N, N*-dicyclohexylcarbodiimide in the presence of the acyl transfer catalyst,

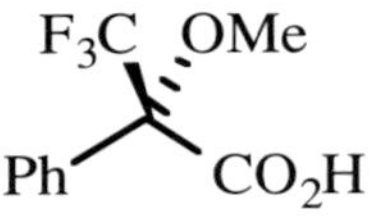

ANALYTICAL METHODS

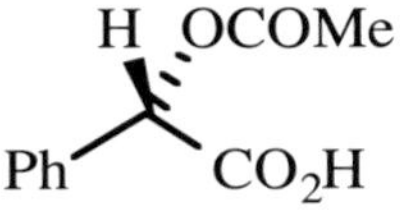

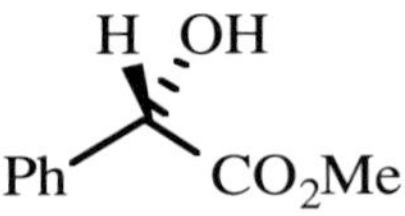

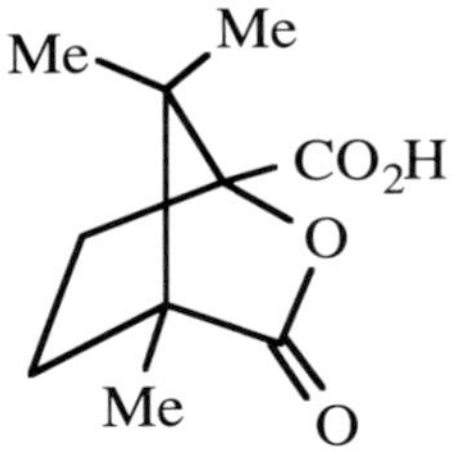

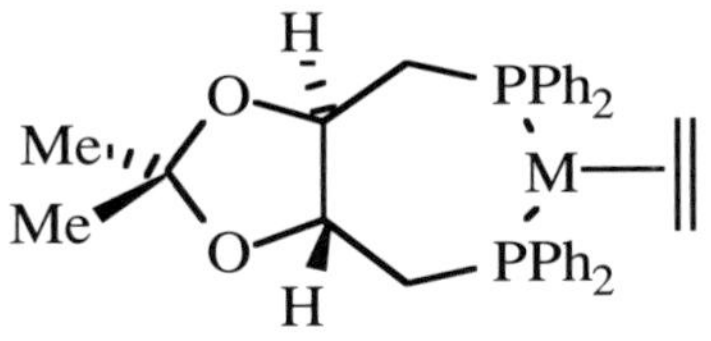

[Structures (10), (11), (12), (13), (14) with (a) M = Pt, (b) M = Pd]

4-dimethyl-aminopyridine (Scheme 3.1). Integration of ^{1}H NMR spectra is reliable for enantiomeric excesses up to 99%. Mandelates are particularly useful in this context because the mandelate CH resonance is not spin coupled and occurs at around 6.0 ppm in a frequently unobscured 'window' in an otherwise crowded proton

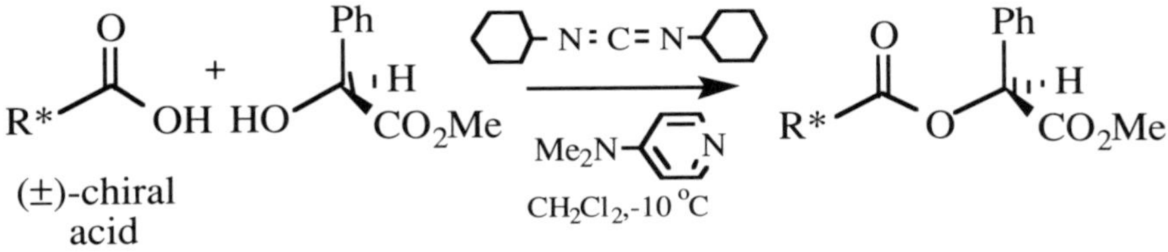

[Scheme 3.1 reaction diagram]

Scheme 3.1 Derivatisation of chiral acids with S-methyl mandelate involving esterification under non-racemising conditions.

NMR spectrum. Anisochronicity ($\Delta\delta \approx 0.2$ ppm, 298 K) is commonly induced in more than one resonance giving an instant internal cross reference (Figure 3.3).

Silyl acetals have been used in the assay of the enantiomeric purity of alcohols. A dichlorosilane is first reacted with an alcohol of 100% enantiomeric excess, such as methylmandelate, quinine or menthol, to give a CDA which is then used to derivatise a second alcohol of unknown enantiomeric purity (Scheme 3.2). ^{1}H, ^{13}C and ^{29}Si NMR

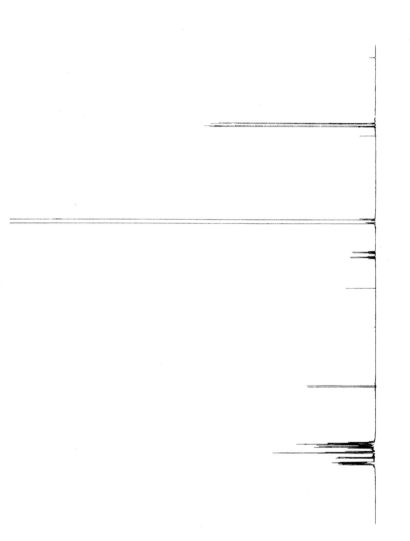

Figure 3.3 500 MHz proton NMR spectrum (d_6-benzene, 298 K) for racemic 2-phenylpropionic acid derivatised with (*S*)-methyl mandelate.

chemical shift non-equivalence has been observed in the diastereomers so formed and agreement between integrated NMR e.e. determinations and optical purities is within 3%.

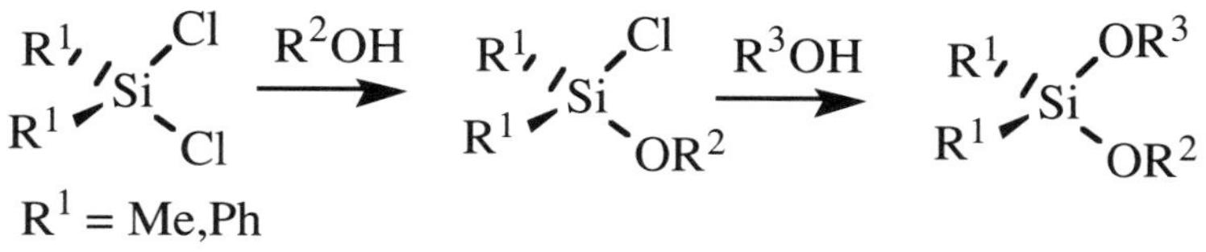

$$R^1_2SiCl_2 \xrightarrow{R^2OH} R^1_2Si(Cl)(OR^2) \xrightarrow{R^3OH} R^1_2Si(OR^3)(OR^2)$$

R^1 = Me, Ph

Scheme 3.2 Reaction of silyl dichlorides with enantiomerically pure alcohol R^2OH to give a CDA which is then used to derivatise the alcohol R^3OH of unknown enantiomeric purity.

^{31}P NMR is a useful alternative to proton NMR. It is a reasonably sensitive nucleus which does not suffer from the relaxation problems associated with ^{13}C NMR. Phosphorus trichloride can be used as a reagent for self-recognition by a chiral substrate. Two molecules of an enantiomerically enriched alcohol react with each PCl_3 molecule to form a phosphonate. Four stereochemically distinct species are possible: (R,R) and (S,S) (a pair of enantiomers), (R,S) and (S,R) which are *meso* compounds. There are consequently three resonances in the ^{31}P NMR spectrum, whose non-equivalence is typically 0.5 ppm and integration gives results within 2% of those obtained by chiral GC methods. Subsequently methyl phosphoryl chloride has been employed in the same way giving improved chemical shift non-equivalences, $\Delta\delta \approx 1$ ppm (Scheme 3.3).

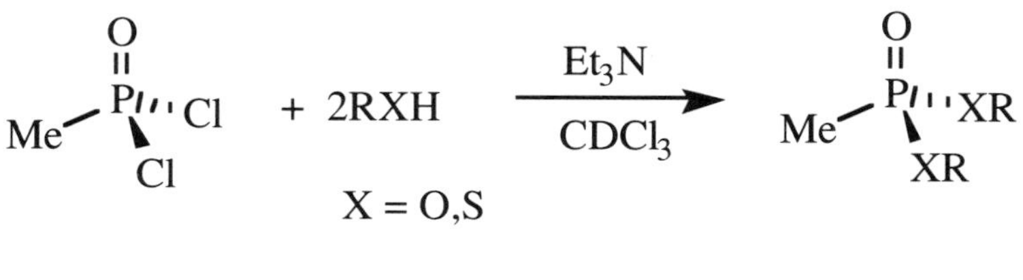

$$\text{MeP(O)Cl}_2 + 2RXH \xrightarrow[CDCl_3]{Et_3N} \text{MeP(O)(XR)}_2$$

X = O, S

Scheme 3.3 Reaction of methylphosphoryl chloride with alcohols or thiols of unknown enantiomeric purity in self-recognition derivatisation.

Recently self-recognition derivatisation has been carried out *in situ* —an approach which combines the convenience of a CSA with the $\Delta\delta$ magnitude advantage of a CDA. A racemic alcohol is derivatised with

a phosphorothioic acid to give all four possible diastereoisomeric
O,O-dialkylphosphorothioate esters (Scheme 3.4).

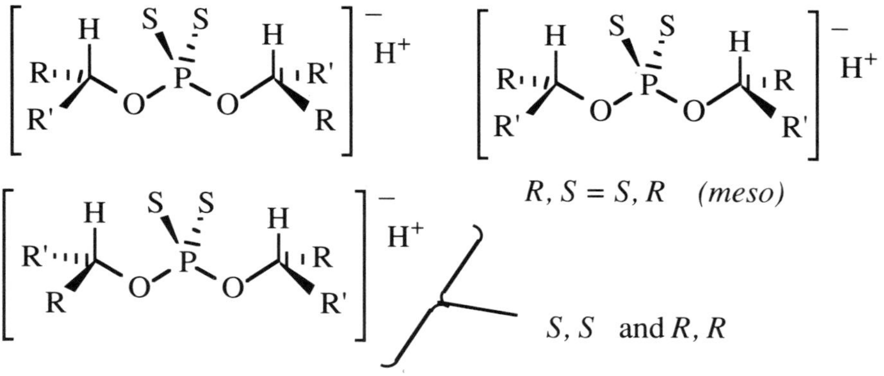

$R, S = S, R$ *(meso)*

S, S and R, R

Scheme 3.4 Three possible diastereomers obtained when the chiral alcohol RR'CHOH is derivatised with a phosphorothioic acid.

The enantiomeric composition of multiply bonded species can be assayed using an organometallic CDA.[10] An enantiomerically pure chiral phosphine, 2,2-dimethyl-4,5-bis(diphenylphosphino)-1,3-dioxolane (diop), is used to prepare a platinum⁰ or palladium⁰ ethene complex, (**14**). *In situ* derivatisation involves the displacement of the ethene ligand by the alkene enantiomers. ^{31}P NMR reveals not only the enantiomeric purity of the alkene substrate but also can assist in the assignment of absolute configuration by comparison of the binding of the *Si* and *Re* faces of the alkene. This technique has found a particular application in determination of the enantiomeric purity achieved in starting material recovered from kinetic resolutions of alkenes using chiral organorhodium catalysts (Figure 3.4).

3.4.2 Chiral solvating agents (CSAs)

When a chiral solute dissolves in a chiral solvent then a stereochemical interaction must be involved. The expense of using a chiral material as a bulk solvent for NMR determination of enantiomeric purity is rarely justified. A solvating agent is added in between 1 to 10 mole equivalents to a solution of the solute enantiomers in an achiral bulk

ANALYTICAL METHODS

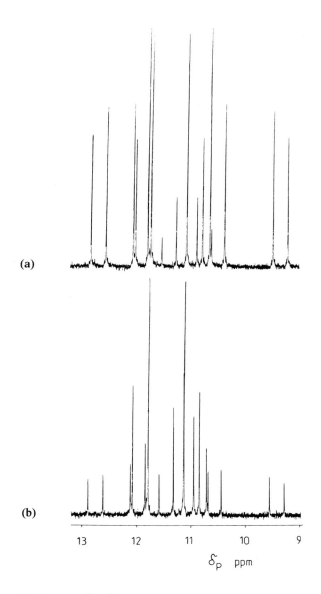

(a)

(b)

$\tilde{\delta}_P$ ppm

Figure 3.4 202 MHz ^{31}P {^{1}H} NMR spectra (d_6-benzene, 298 K) for enriched mixtures of dimethyl 3-methylitaconate derivatised with (**14a**). A : Itaconate recovered after 46% hydrogenation with (S)-BINAP rhodium catalyst, e.e. ≈ 32% in favour of (R)-enantiomer. B: Itaconate recovered after 25% hydrogenation with (R, R)-DiPAMP rhodium catalyst, e.e. ≈ 31% in favour of (S)-enantiomer.

solvent. Chiral solvating agents form diastereomeric solvation complexes with the substrate enantiomers *via* rapidly reversible equilibria in competition with the bulk solvent.

Chemical shift anisochronicity in the presence of a CSA has two origins. The first is the stereochemically dependent position of groups with large magnetic anisotropy, such as phenyl rings and carbonyl functions, relative to other substituents in each of the solvate complexes. The second is the relative magnitude of the solvation constants K_R and K_S for the solute enantiomers. Exchange between the chiral and achiral solvates is rapid on the NMR timescale. The observed resonance signals for each enantiomer, $\delta_R(\text{obs})$ and $\delta_S(\text{obs})$ respectively, comprise population weighted averages of the chemical shifts for the discrete chiral and achiral solvates δ_R, δ_S and $\delta_\pm$

$$\delta_R(\text{obs}) = \phi_R.\delta_\pm + (1 - \phi_R).\delta_R$$
$$\delta_S(\text{obs}) = \phi_S.\delta_\pm + (1 - \phi_S).\delta_S$$

where ϕ_R and ϕ_S are the fractional populations of achiral solvates.

$$K_R = (1 - \phi_R)/\phi_R \text{ and } K_S = (1 - \phi_S)/\phi_S$$
$$\Delta\delta = \phi_R.(\delta_\pm + K_R.\delta_R) - \phi_S.(\delta_\pm + K_S.\delta_S)$$

The advantage of the CSA technique is that it is quick and simple, requiring no separate derivatising reaction prior to NMR assay. There is no problem with accidental enrichment or racemisation of the sample due to differential reaction rates, provided that the sample remains in solution in the presence of the CSA. The enantiomeric purity of the CSA is not critical. If a CSA of less than 100% e.e. is used, the magnitude of the chemical shift non-equivalence is reduced but the relative signal intensities are not affected. This is a consequence of the NMR time-averaged view of the rapid exchange processes involved and is in contrast to the CDA technique. Solvation models permit the sense of non-equivalence to be correlated with the absolute configuration of the solute.

The magnitude of the chemical shift non-equivalence is generally smaller when using CSAs in comparison with CDAs. Non-polar achiral solvents maximise the anisochronicity between the diastereomeric complexes while polar solvents effectively exclude formation of these complexes with the CSA and hence reduce $\Delta\delta$ to zero.

The most commonly used CSAs are a series of 1-aryl-2,2-2-trifluoroethanols (**15**). It is generally accepted that for induction of chemical shift non-equivalence each solvation complex must feature a minimum of three interactions. Two interactions are necessary

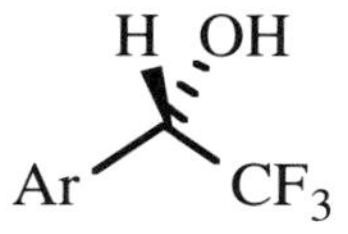

(**15**) (**a**) Ar = 9-anthryl
(**b**) Ar = 10-methyl-9-anthryl
(**c**) Ar = phenyl
(**d**) Ar = cyclohexyl

to form a 'chelate-like' structure, the third interaction must be stereochemically dependent and is responsible for causing anisochronicity in solute substituents. Thus, the CSA and the solute must have complementary functionality if the necessary interactions are to occur. These can be hydrogen bonding, charge transfer (π-acid–π-base), dipole–dipole, and proton transfer.

In Scheme 3.5 the relatively acidic hydroxyl and methine protons are involved in hydrogen-bonding with the primary and secondary basic sites in the solvate, B^1 and B^2, to give complexes I and II. The solvate substituents R^1 and R^2 experience differential shielding due to the CSA aryl substituent. The opposite sense of non-equivalence for substituents on opposing sides of the chelate plane is a hallmark of the CSA technique.

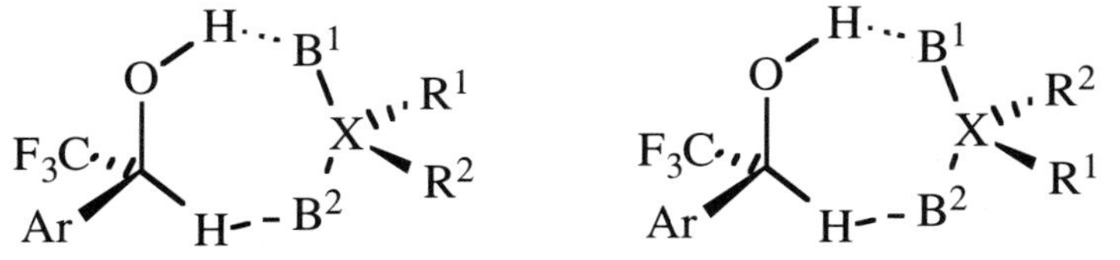

Scheme 3.5 Binding models proposed to explain the induction of chemical shift non-equivalence by 1-aryl-2,2,2-trifluoroethanol CSAs.

Distinct ^{19}F NMR resonances were first observed for the enantiomers of 2,2,2-trifluoro-1-phenylethanol in the presence of (R)-phenylethylamine. With (R)-2-naphthylethylamine the magnitude of the non-equivalence was increased. A systematic study of a series of aryl alcohols in the presence of amines showed a consistent correlation between the sense of non-equivalence and the absolute configuration of the alcohol. From the simple solvation models proposed, the 'reciprocality' of the CSA approach is evident, i.e. if chiral A can be used to assay racemic B then chiral B can be used to assay racemic A. With this in mind 1-(9-anthryl)-2,2,2-trifluoroethanol (**15a**) was developed as a CSA for chiral amines. It is also effective with alcohols, lactones, α-amino acid esters, α-hydroxy acid esters and sulphoxides and is the most widely used chiral solvating agent. Other more specific solvating agents have been developed. Kagan has developed N-(3,5-dinitrobenzoyl)-1-phenylethylamine, for example, as a CSA specifically for the assay of chiral sulphoxides prepared from sulphides by a modified Sharpless oxidation (section 6.3.2).

The solvation models proposed by Pirkle are clearly an over-simplification in many cases. For example, with methyl isopropyl sulphoxide in the presence of 1-(9-anthryl)-2,2,2-trifluoroethanol, $\Delta\delta$ continues to increase after one equivalent of the solvating agent has been added and reaches a maximum with three equivalents of CSA.

Changing the temperature at which the NMR spectrum is recorded has a marked effect on the magnitude of the chemical shift non-equivalence. A drop in temperature of, say, 25°C can double $\Delta\delta$ as stereochemically important conformations become preferentially populated (Figure 3.5). In addition, careful study of the relative shifts of resonances with temperature variation can give insights into the structure of the solvation complexes.

Diastereomeric salt formation results from complete proton transfer between CSA and solute. There is rapid exchange between the free acid and base and that constituting a close ion-pair. Solvation of such systems can be problematic since by its nature a salt requires a reasonably polar solvent. However, such solvents tend to dissociate the close ion-pairs to give solvent-separated ion-pairs in which the stereochemically dependent interaction responsible for induction of anisochronicity is lost. This limitation can be overcome in some cases by using mixed achiral solvents such as d_6-benzene and d_5-pyridine.

No models have been proposed to account for the observed $\Delta\delta$

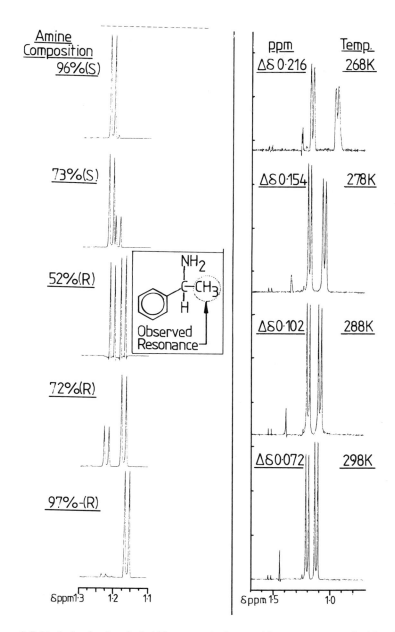

Figure 3.5 Variation in chemical shift non-equivalence with temperature and with solute enantiomeric composition. Solute: 1-phenylethylamine. CSA: (*S*)-*O*-acetylmandelic acid. Proton NMR spectra recorded at 250 MHz in d-chloroform.

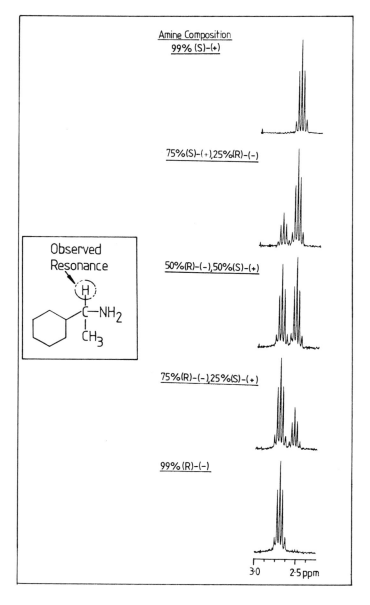

Figure 3.6 250 MHz proton NMR spectra (d-chloroform, 298 K) for mixtures of 1-cyclohexylethylamine of varying enantiomeric composition in the presence of (S)-O-acetylmandelic acid as a CSA.

which will depend on the size of any ionic aggregates in the solution and on the relative values of the four equilibrium constants. CSA-solute interactions of this type are limited to systems capable of salt formation. Mosher's MTPA[9] and Parker's O-acetylmandelic acid[11] have both been used to assay the enantiomeric purity of all classes of amines (Figure 3.5) and, in the latter case, β-amino-alcohols such as propranalol and salbutamol which had previously been inaccessible due to solubility problems. The magnitude of chemical shift non-equivalence can vary not only with temperature but also with solute enantiomeric composition (Figure 3.6). If resolution is significantly reduced at high solute e.e. then use of the opposite enantiomer of the CSA solves the problem. In contrast with some other systems, $\Delta\delta$ is a maximum when the CSA:solute ratio is 1:1, indicating that a stoichiometric salt is formed. Applying the reciprocality principle for CSA–solute interactions, chiral amines are expected to behave as CSAs for chiral acids. Camphanic acid has been assayed, for example, using (R)-phenylethylamine. Chiral solvating agents offer the most elegant solution to the determination of enantiomeric purity using NMR. The full potential of the technique has yet to be realised, the main drawback being that, as with other resolving agents, there is no 'universal agent' applicable to all systems. The best that can be achieved at present is to make an informed choice of CSA according to the specific problem in hand. The main advantage of CSAs over CDAs is their ease of use which allows the researcher to identify quickly the CSA appropriate to the problem. The chief disadvantage is the smaller $\Delta\delta$ induced, although the advent of very high field NMR spectrometers has gone some way to assuage this difficulty.

3.4.3 Chiral lanthanide shift reagents (CLSRs)

Hinckley demonstrated in 1969[12] that paramagnetic tris-(β-ketonato)-lanthanide(III) chelates are capable of inducing large shifts in certain resonances of the NMR spectra of organic substrates. The most widely used achiral lanthanide shift reagents (LSRs) are tris-(2,2,6,6-tetramethylheptane-3,5-dionato)-europium(III) [Eu(thd)$_3$] (**16**) and tris-1,1,1,2,2,3,3-heptafluoro-7,7-dimethyloctane-1,6-dionato)-europium(III) [Eu(fod)$_3$] (**17**), and the corresponding praseodymium and ytterbium complexes. These are used primarily to simplify proton NMR spectra by shifting resonances to less congested areas.

(16), (17) [structures: tris-chelate Eu complexes with t-Bu/t-Bu and t-Bu/CF₂CF₂CF₃ β-diketonate ligands]

The induction of resonance shifts, denoted $\Delta\delta$, depends on the establishment of a fast reversible molecular association equilibrium between the LSR and the organic donor substrate. The LSR is able to expand its coordination number beyond six by accepting electron density from donor functionalities in the substrate such as amino, hydroxy or carbonyl groups. The f-shell electrons of the rare earth metal are not available for covalent bonding and consequently induced resonance shifts result only from dipolar interactions through space, sometimes known as pseudo-contact interactions. The magnitude of the pseudo-contact shift decreases with the cube of the distance of the nucleus from the interaction site, thus LSRs can give information about the molecular conformation of the substrate. Chiral lanthanide shift reagents (CLSRs), such as tris(t-butyl-hydroxy-methylene-(1R)-camphorato)europium(III) (18), form diastereomeric association complexes with substrate enantiomers in solution which are subject to similar equilibria to CSA complexes. The induced shift in any

(18) R = Bu^t
(19) R = CF_3
(20) R = $CF_2CF_2CF_3$

one substrate resonance will be different in each diastereomeric complex. In the ^1H NMR spectrum, therefore, there will be an induced shift non-equivalence, $\Delta\Delta\delta$. This effect was first observed in the ^1H NMR spectrum of racemic 1-phenylethylamine in the presence of (**18**). The induced high frequency shift of the amine methine proton was 17 ppm. The chemical shift non-equivalence was 0.05 ppm (in CCl_4, 298 K). The perfluoroalkyl camphorato chelates tris-(3-trifluoromethyl-hydroxymethylene-(1R)-camphorato)europium(III) [Eu(thd$_3$)] (**19**), and tris-(3-heptafluorobutyryl-hydroxymethylene-(1R)-camphorato)-europium(III) [Eu(hfc)$_3$] (**20**), have been used extensively to measure the enantiomeric purity of chiral alcohols, aldehydes, ketones, esters, oxiranes and sulphonamides.[13] Surprisingly, there are few reports of the use of carboxamides as substrates notwithstanding their ability to bind strongly to europium. The combination of Eu(hfc)$_3$ with the achiral β-diketonate complex Ag(fod) permits the assay of the enantiomeric purity of a limited range of chiral alkenes.

The mechanism of chemical shift non-equivalence induction is more complicated in the case of CLSRs than with CSAs. The shift reagents themselves can exist as a number of rapidly interconverting isomers. Frequently, several equivalents of CLSR must be added to a solution of the chiral substrate in a bulk solvent before anisochronicity is observed. The precise nature of the association complexes formed varies from one example to the next. This makes assignment of absolute configuration on the basis of the sense of non-equivalence unreliable even in structurally related series of compounds.

The association complexes formed are especially moisture sensitive; the CLSR and the substrate must be dry and the solvent should be dried over molecular sieve immediately prior to use. The optium experimental conditions are achieved when $\Delta\Delta\delta$ is maximised and the concomitant line broadening minimised. Line broadening is caused by the slowing of the dynamic equilibria to a rate comparable to that of the NMR timescale. In practice, the CLSR is added starting at 0.5 molar equivalents, and increments of 0.5 molar equivalents are added sequentially until resolvable anisochronicity is observed. Although the chemical shift non-equivalence decreases with decreasing CLSR enantiomeric purity, the relative areas under the resonance peaks corresponding to the diastereoisomeric complexes do not alter. Thus the measurement of substrate enantiomeric purity by integration is not impaired by low CLSR e.e. provided that there is still baseline resolution of the appropriate resonances.

The magnitude of the chemical shift non-equivalence depends on the strength of the complexation, the proximity of the stereogenic centre to the complexation site and the identity of the substrate functionalities involved in electron donation. Non-polar solvents such as carbon tetrachloride or d-chloroform are preferable, although the europium(III)-(R)-propylenediaminetetra-acetate ion has been used in D_2O to assay unsubstituted acids. The effect of lowering the temperature can be advantageous in the case of weakly complexed substrates or those which owe their chirality to only small differences in substituents. The enantiomers of 2-methyl-1-butanol, 1-nitroethanol and 2-cyanobutane were resolved at 243 K where anisochronicity was not observed at 298 K. However, lowering the temperature can cause problems with line broadening.

Reported CLSR enantiomeric purity determinations are in good agreement with the results from other methods (±2% in 30%). The best claimed results are ±2% in the range 40–60% e.e. but it has been claimed that for e.e. ≥90% the error can be as large as 10%.

A practical example of the use of this technique is shown in Figure 3.7. The tricyclic compound (**21**) was recently prepared in high e.e. by cinchona alkaloid catalysed ring-opening of the epoxyanhydride with methanol.[14] Its e.e. was determined using the CLSR (**20**) by integrating the sharp singlets at δ4.2 due to the methyl group. Several

features of the spectra are notable: all signals are shifted to higher frequency on adding the shift reagent but those due to protons nearest the presumed binding site for Eu (the OH group) are shifted most. Several of the signals are split, but it is only the sharp singlet due to the methyl group for which baseline separation is achieved. The broad signals at δ4.4, 0.3 and 0.2 are due to the shift reagent itself.

3.5 Concluding remarks

The modern chemist engaged in the determination of enantiomeric purity needs to be aware of the breadth of techniques that are available. It should become standard practice for workers to determine

ANALYTICAL METHODS 61

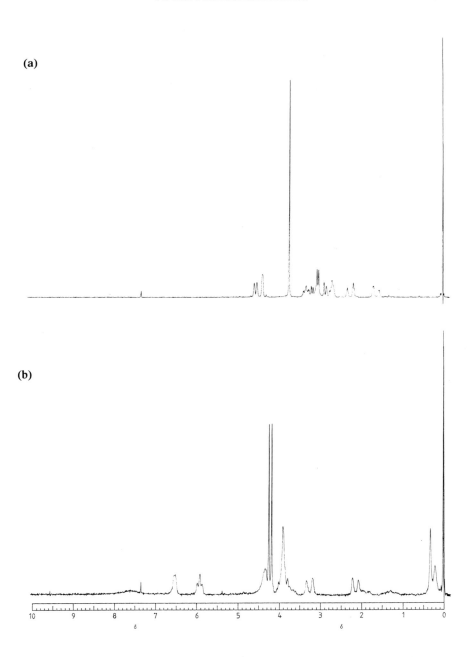

Figure 3.7 80 MHz proton NMR spectra (d-chloroform, 298 K) of (a) compound (**21**) and (b) racemic (**21**) in the presence of Eu(hfc)$_3$ (**20**) as CSR.

enantiomeric purities by at least two independent methods. The limitations of the traditional polarimetric means of analysis—including the idea that optical purity and enantiomeric purity are not necessarily equivalent—must be realised. With the heightened interest in highly enantioselective reactions (e.e. ≥ 95%), and the more stringent demands for licensing pharmaceuticals of precisely known enantiomeric composition, it is particularly important to have accurate and sensitive methods of analysis. Furthermore the myths that naturally occurring compounds or commercially available chiral auxiliaries are enantiomerically pure must be dispelled.

Of the modern methods of analysis, chiral GC and HPLC methods using a chiral stationary phase are probably the most sensitive, detecting as little as 0.1% of one enantiomer in the presence of another. They still tend to require pre-column derivatisation to enhance separability or to improve the ease of detection: this extra step may inhibit the general acceptance of the methods unless automated procedures are developed. Furthermore, the available chiral stationary phases are somewhat restricted in the types of compound for which analysis is feasible. The modern NMR spectroscopist has three techniques in his armoury with which to tackle enantiomeric purity analysis. Chiral derivatising agents are reliable and effective but experimentally inconvenient while chiral lanthanide shift reagents are equally effective but their application needs a trial and error approach to optimise conditions for a specific system. Chiral solvating agents are elegant and simple to use but are perhaps under-developed and hence so far are restricted in their range of applications.

References and notes

1. A. Horeau and J.P. Guette, *Tetrahedron,* 1974, **30**, 1923.
2. P. Crabbé, *Optical Rotatory Dispersion and Circular Dichroism in Organic Chemistry*, Holden-Day, San Francisco, 1985.
3. A useful review of this method has appeared: V Schürig and A. P. Novotny, *Angew. Chem. Int. Ed. Engl.*, 1990, **29**, 939.
4. (a) J. Lough (ed), *Chiral Chromatography*, Blackie, Glasgow, 1989; (b) W.H. Pirkle and J. Finn, in *Asymmetric Synthesis*, Vol. 1, J.D. Morrison, ed., Academic Press, Orlando, 1983, Chapter 6, p. 87.
5. S. Yamaguchi in reference [3a], Chapter 7.
6. G.R. Weissman in reference [3a], Chapter 8.

7. R.R. Fraser in reference [3a], Chapter 9.
8. K. Mislow and M. Raban, *Tetrahedron Lett.*, 1965, 4249.
9. J.A. Dale, H.S. Mosher and D.L. Dull, *J. Org. Chem.*, 1969, **34**, 2543.
10. D. Parker and R.J. Taylor, *J. Chem. Soc. , Chem. Commun.*, 1987, 1781; *Tetrahdron* , 1988, **44**, 2241.
11. D. Parker and R.J. Taylor, *Tetrahedron* , 1987, **43**, 5451.
12. C.C. Hinckley, *J. Am. Chem. Soc.*, 1969, **91**, 5160.
13. G.R. Sullivan, *Top. Stereochem.*, 1979, **10**, 287.
14. R.A. Aitken, J. Gopal and J.A. Hirst, *J. Chem. Soc., Chem. Commun.*, 1988, 632.

4 Sources and strategies for the formation of chiral compounds
R. A. AITKEN and J. GOPAL

The ultimate source of chirality in all asymmetric synthesis is nature. In this chapter we survey the most important naturally occurring chiral compounds and the ways in which they can be used. This is followed by a classification of the known methods of asymmetric synthesis and a general consideration of their mechanisms.

4.1 Chiral starting materials

The chiral compounds which occur in nature provide an enormous range and diversity of possible starting materials. To be useful in asymmetric synthesis, these should be cheap and readily available in high enantiomeric purity. For many applications the availability of both enantiomers is advantageous. Most importantly, they must be capable of exerting a high degree of stereocontrol in the required reactions by means of steric hindrance, chelation or other specific effects. An extensive list of such compounds together with an indication of cost has been compiled.[1] In the following sections the major classes of chiral starting materials are examined.

4.1.1 Amino acids and amino alcohols

The twenty common α-amino acids which make up proteins, excluding the achiral glycine, are perhaps the most widely used single class of chiral starting materials for asymmetric synthesis.[2] The amino acid functionality is particularly versatile and can be used in many ways. The amino alcohols readily derived by reduction of the carboxyl group have also been much used and, when they are attached by the amino group, the alcohol OH often has an important chelating function. In many asymmetric reactions the stereocontrol is achieved by steric effects and so amino acids with bulky side-chains have proved particularly useful. Examples are provided by (S)-valine (**1**) and (S)-phenylalanine (**2**) together with the corresponding amino alcohols (S)-valinol (**3**) and (S)-phenylalaninol (**4**) and the compound (**5**) in which the bulk has been further increased by formation of the *t*-butyl ether.

The amino acids are obtained either from protein hydrolysis or by microbiological processes in high enantiomeric purity in favour of the (S)-enantiomer [(R) for cysteine]. For this reason the opposite unnatural enantiomers are always more expensive, sometimes very much more so. Thus, while (R)-asparagine costs only twice as much as the (S)-enantiomer, the ratio in the case of arginine is about 400:1. The reason for wanting both enantiomers, of course, is that in very many processes their use will give access to opposite enantiomers of the product. The extra degree of rigidity in the cyclic amino acid (S)-proline (**6**) makes it particularly useful in directing many asymmetric processes but unfortunately (R)-proline is very difficult to obtain from natural sources. This problem has been overcome by using (R)-pyroglutamic acid (**7**), readily derived from the inexpensive (R)-glutamic acid as an indirect source.

While the amino acids with hydrocarbon side-chains (alanine, valine, leucine, isoleucine, phenylalanine, phenylglycine) are used largely for their steric directing effects, others with functional groups, particularly OH (serine, threonine), NH_2 (lysine, ornithine), SH (cysteine), CO_2H (aspartic, glutamic acids) and $CONH_2$ (asparagine, glutamine) allow incorporation of the amino acid for use in a wide variety of applications using the side-chain functionality in addition to the amino acid. It is important to realise that these amino acids, and indeed many of the chiral compounds in this section, are only the primary source of chirality and they may need to undergo several steps for conversion into a form suitable for use as a chiral auxiliary, reagent or catalyst.

In addition to the amino acid derived amino alcohols mentioned above, two other amino alcohols have been widely used: (1R, 2S)-ephedrine (**8**) which is a naturally occurring plant alkaloid, and the (S, S)-amino diol (**9**) which is produced microbiologically as a by-product from the manufacture of the antibiotic chloramphenicol. Note the presence of two adjacent [or *contiguous*] stereogenic centres in both these molecules, a useful feature also present in two of the amino acids, (S, S)-isoleucine (**10**) and (2S, 3R)-threonine (**11**). The (R)-amino alcohol (**12**) is also readily available in enantiomerically pure form.

4.1.2 Hydroxy acids

The α-hydroxy acids represent another widely used class of chiral starting materials. Some, such as (S)-lactic acid (**13**) and (S)-malic acid (**14**), are readily available from natural sources while others, such as (R)-mandelic acid (**15**) and its enantiomer, can be obtained by resolution of the racemic synthetic material. The useful arrangement of functional groups in the dihydroxy diacid tartaric acid has proved invaluable in certain asymmetric reactions in which it acts as a bidentate ligand or auxiliary group. Both the (R, R)-isomer (**16**) and the slightly more expensive (S, S)-isomer (**17**) are readily available. Other hydroxy acids can be obtained from more unusual sources. A good example is provided by the enantiomers of 3-hydroxybutyric acid. Under appropriate conditions certain bacteria can produce and store poly(3-(R) hydroxybutyrate) which may make up as much as 80% of their dry weight. Simple acid hydrolysis of such a culture affords the 3-(R)-hydroxybutyrates (**18**) in very high e.e. On the other hand 3-(S)-hydroxybutyrates (**19**) are readily available from reduction of the corresponding acetoacetates with baker's yeast.

4.1.3 Alkaloids and other amines

Many plant species produce complex nitrogen-containing secondary metabolites which were first investigated because of their physiological activity. These alkaloids, which generally occur in enantiomerically pure form, have been used since the last century for resolution of acids by the formation of diastereomeric salts. More recently their potential as catalysts for asymmetric synthesis has been recognised and the rigid three-dimensional arrangement of functional groups present has in some cases given almost enzyme-like selectivity.[3] The cinchona alkaloids quinine, quinidine, cinchonine and cinchonidine have perhaps been the most useful: they are readily available because of the commercial cultivation of the *cinchona* tree to obtain quinine for medicinal purposes, they do not suffer from the problem of extreme toxicity associated with other alkaloids classically used for resolution such as strychnine and brucine (see section 4.2.2), and the combination of the quinoline, tertiary amine, alcohol and vinyl groups offers many possibilities for their exploitation to control asymmetric reactions. Other alkaloid types such as (+)-sparteine (**20**) have also proved useful.

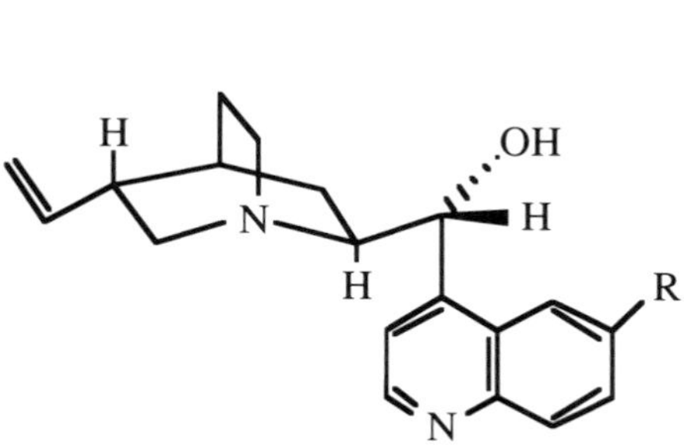

Quinine (R = OMe)
Cinchonidine (R = H)

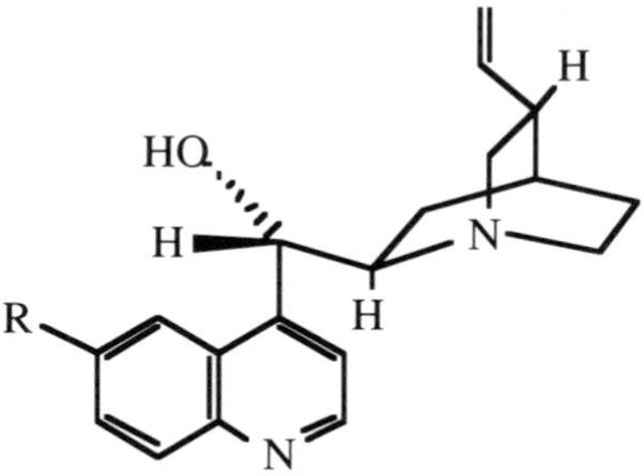

Quinidine (R = OMe)
Cinchonine (R = H)

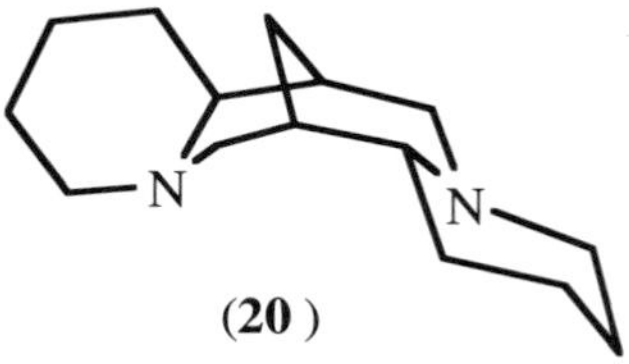

(20)

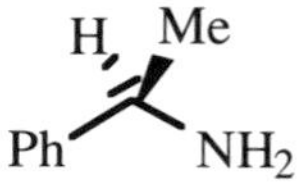

(21)

A number of simpler chiral amines are also available in high enantiomeric purity from resolution. Both (*R*)-1-phenylethylamine (or α-methylbenzylamine) (**21**) and its (*S*)-enantiomer can be obtained at reasonable cost and have been widely used.

4.1.4 Terpenes

Apart from certain carbohydrates, the most inexpensive source of chiral compounds is the terpenes. These are readily obtained from plant sources and encompass examples of many important functional groups. These include alcohols such as (+)-menthol (**22**) and (–)-borneol (**23**), ketones such as (+)-camphor (**24**), (+)-pulegone (**25**), (–)-menthone (**26**) and (–)-carvone (**27**), the aldehyde (+)-citronellal (**28**), (+)-camphor-10-sulphonic acid (**29**), and alkenes such as (+)-limonene (**30**) and (+)-α-pinene (**31**). (α)-Pinene provides a good illustration of the fact that naturally derived chiral compounds are not necessarily enantiomerically pure. Both enantiomers are readily available but the normal samples are only of around 90% e.e. Fortunately this is not a serious problem since procedures have been developed for enantiomeric enrichment of the chiral reagents derived from them (see section 6.4.3).

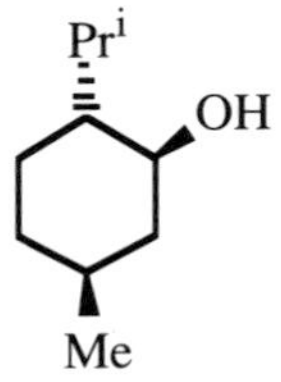

(22) 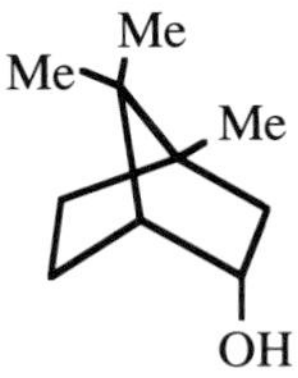 (23) 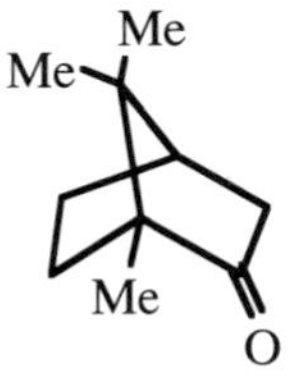 (24) 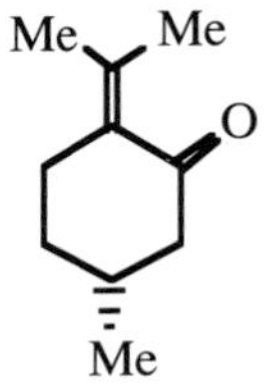 (25) 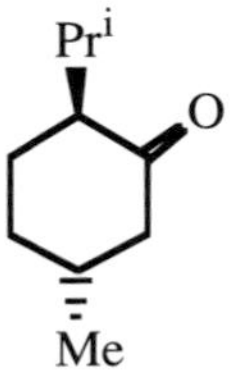 (26)

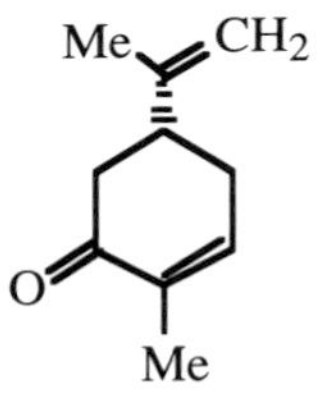

(27) 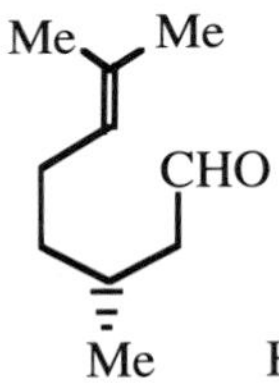 (28) 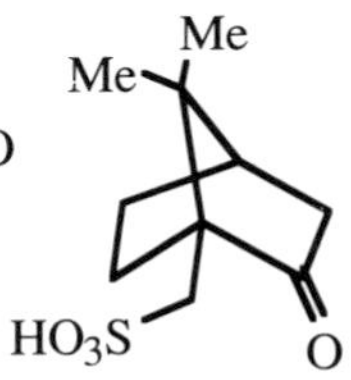 (29) 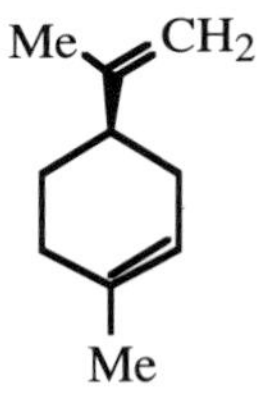 (30) 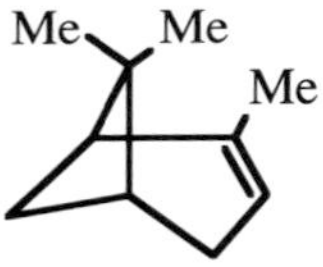 (31)

4.1.5 Carbohydrates

The carbohydrates—sugars and related polyhydroxy compounds—include the most inexpensive of all chiral compounds. With up to four contiguous stereogenic centres they might seem to be extremely valuable starting materials for asymmetric synthesis. The problem arises from the presence of several very similar CHOH groups which can only be chemically differentiated by the careful use of protecting groups. In a sense most carbohydrates have too many stereogenic centres to be easily used as directing groups for asymmetric synthesis. While some methods using carbohydrates in this way have been developed, their most important use is as a source of chiral building blocks for direct incorporation into the structures of target molecules.[4] Quite often this requires the selective destruction of several of the original stereogenic centres.

70 ASYMMETRIC SYNTHESIS

An important building block (**33**) is obtained in an interesting process from the chiral carbohydrate derivative (+)-mannitol (**32**), which has a C_2 axis of symmetry, but no mirror plane, through the central bond. Cleavage of the central bond of the diacetonide with $NaIO_4$ gives two identical molecules of the chiral glyceraldehyde acetonide (**33**). The use of this chiral starting material to prepare (–)-prostaglandin E_1 is described in section 7.2.1.

4.2 Methods for the formation of chiral compounds

Before considering in detail the different approaches to asymmetric synthesis, it is worth looking briefly at all the methods available to obtain chiral compounds in non-racemic form. Some methods, such as the spontaneous crystallisation of enantiomers and 'absolute asymmetric synthesis' using circularly polarised light, are of little general value and will not be considered here.

4.2.1 Use of naturally occurring chiral compounds as building blocks

Perhaps the most obvious way to obtain a chiral compound is simply to take an appropriate chiral precursor from a natural source and modify the structure in a series of chemical steps to arrive at the desired compound. The steps will generally not involve the stereogenic centre(s) required for the product and indeed care has to be taken that there is no chance of racemisation during the sequence. In some cases a stereogenic centre may undergo reaction as long as this proceeds with retention of chirality, an S_N2 substitution being a common example. It may also be necessary to carry out steps which destroy

SOURCES AND STRATEGIES

some of the original stereogenic units It is easy to recognise this type of approach to chiral compounds since *no new stereogenic units are formed*. The stereogenic unit(s) of the product are all directly derived from the starting material.

Although this method might be considered the least elegant approach to chiral compounds, and is in no sense an asymmetric synthesis, methods of this type are of the greatest importance in some areas. For example, although elegant asymmetric total syntheses have been described for very many complex natural and biologically active products, those which are required for use on a large scale are almost invariably made by starting with a chiral precursor from a natural source where this is possible. The preparation of steroids from hecogenin (**34**) which is obtained from the sisal plant is a good example—the formation of each of the stereogenic centres in a steroid such as cortisone (**35**) from scratch using the methods of asymmetric synthesis would clearly be uneconomic and would never be considered when such a convenient chiral starting material is available. Some examples of this type of synthesis are described in section 5.1.

4.2.2 Resolution

Resolution may be considered the classical method of obtaining enantiomerically pure products. The procedure relies on the fact that diastereomers, unlike enantiomers, have different physical properties. If the racemic compound which is to be resolved is derivatised by reaction with a naturally occurring enantiomerically pure compound, then the resulting diastereomeric compounds may be separated, most commonly by crystallisation but also by chromatography, and then separately treated to liberate the two enantiomers. If we represent the substrate to be resolved by S and the resolving agent by A*, then the overall process is shown by:

$$(\pm)\text{-S} \xrightarrow{+A^*} (+)\text{-S.A}^* + (-)\text{-S.A}^* \quad \text{(diastereomers)}$$

$$\downarrow \text{separate}$$

$$(+)\text{-S} \xleftarrow{-A^*} (+)\text{-S.A}^* \quad (-)\text{-S.A}^* \xrightarrow{-A^*} (-)\text{-S}$$

Note that the resolving agent is recovered unchanged after this procedure and can be reused repeatedly. Because of the need to obtain crystalline adducts which are readily broken down to their components again, the ionic salts formed between amines and acids, either carboxylic or sulphonic, are ideal for resolution. Thus even in the last century very many amines were resolved by formation of salts with, for example, tartaric acid (**16**) or camphorsulphonic acid (**29**), while organic acids were resolved with bases such as quinine, cinchonine and the highly toxic alkaloids brucine (**36**) and strychnine (**37**). Although reliable resolution methods have now been worked out for

(**36**) R = OMe
(**37**) R = H

many other types of compound[5] and these often provide the most cost-effective way to obtain enantiomerically pure compounds on a large scale, development of reliable resolution procedures for a given compound is still very much a matter of trial and error. Some examples of syntheses based on resolution are described in section 5.1.

4.2.3 Methods of asymmetric synthesis

As mentioned in chapter 1, asymmetric synthesis involves the formation of a new stereogenic unit in the substrate under the influence of a chiral group ultimately derived from a naturally occurring chiral compound. The known methods can be conveniently divided into four major classes, depending on how this influence is exerted, as follows:

(a) First-generation or substrate-controlled methods

In these, reaction is directed intramolecularly by a stereogenic unit already present in the chiral substrate. The formation of the new stereogenic unit most often occurs by reaction with an achiral reagent at a diastereotopic site controlled by a nearby stereogenic unit. If we represent the part of the substrate which reacts as S, the chiral directing group as G, the reagent as R, the product as P–G and chirality by *, the overall process becomes:

$$\text{S–G}^* \xrightarrow{\text{R}} \text{P}^*\text{–G}^*$$

A specific example is provided by the addition of methyl Grignard reagent to (S)-2-methylcyclohexanone (**38**) to give (**39**) in which addition to the carbonyl group is influenced by the adjacent stereogenic centre according to Cram's rule.

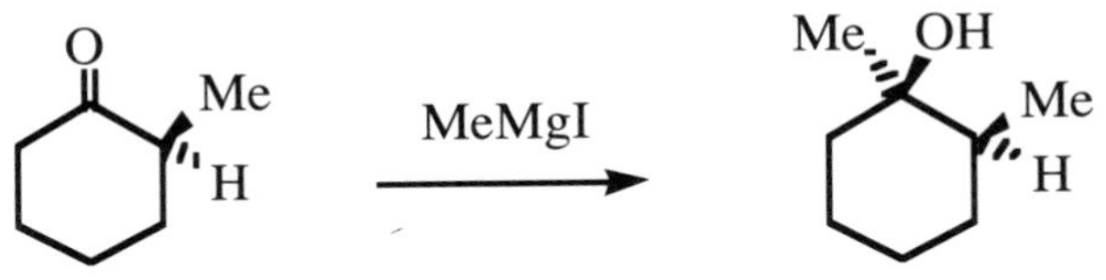

(**38**) $\xrightarrow{\text{MeMgI}}$ (**39**)

The main drawback of this procedure is the need for an enantiomerically pure starting material; we are not forming a chiral product from an achiral substrate but merely adding an additional stereogenic unit to an already enantiomerically pure substrate. Unless the right substrate happens to be readily available, this is obviously of limited value although, as will be seen in chapter 7, many real synthetic schemes involve a second-, third- or fourth-generation method followed by several first-generation steps to introduce the remaining stereogenic units under the influence of those already present. First-generation methods are described in more detail in section 5.2.

(b) Second-generation or auxiliary-controlled methods

This approach is similar to the first-generation method in that control is again achieved intramolecularly by a chiral group in the substrate. The difference is that the directing group, the 'chiral auxiliary', is now deliberately attached to an achiral substrate in order to direct the reaction and can be removed once it has served its purpose. In this way we can achieve overall conversion of an achiral substrate to a chiral product. Retaining the same symbols as above and representing the auxiliary by A, we have:

$$S \xrightarrow{+A^*} S\text{-}A^* \xrightarrow{R} P^*\text{-}A^* \xrightarrow{-A^*} P^*$$

(recycle A^*)

An additional useful feature of this approach is that the two possible products resulting from the alternative modes of reaction with R are not enantiomers but diastereomers as a result of the presence of the additional stereogenic centre of the auxiliary. This means that even if the diastereoselectivity of the reaction is only moderate, the undesired diastereomer of the initial product can be removed by crystallisation or chromatography so that, after removal of the auxiliary, the final product is obtained in very high e.e. An example is the methylation of cyclohexanone, *via* its imine formed with the methyl ether of (*S*)-phenylalaninol (**4**), to give (**38**) as shown:

SOURCES AND STRATEGIES 75

[Scheme showing cyclohexanone + H₂N-CH(CH₂Ph)-CH₂-OMe (−H₂O) → imine; then 1. B⁻, 2. MeI → alkylated imine; then H⁺, PhCH₂-CH(NH₂)-CH₂-OMe → 2-methylcyclohexanone (**38**)]

Most of the new asymmetric synthesis methods introduced in the last 20 years are of the second-generation type and a detailed description of these forms the subject of sections 5.3–5.5.

(c) Third-generation or reagent-controlled methods

Although the second-generation methods have proved very useful, the need for two extra steps to attach and then remove the chiral auxiliary is an unattractive feature. This can be avoided by using a third-generation method in which an achiral substrate is directly converted to the chiral product by the use of a *chiral reagent*:

$$S \xrightarrow{R^*} P^*$$

In contrast to the first- and second-generation methods, the control is now intermolecular. This is obviously an attractive procedure but the range of reactions for which effective chiral reagents exist is somewhat limited at present. An example is provided by the hydroboration of 1-methylcyclohexene, using isopinocampheyl-borane (**40**) derived from (+)-α-pinene (**31**), to give alcohol (**41**) with two adjacent stereogenic centres.

(d) Fourth-generation or catalyst-controlled methods

In each of the previously mentioned three classes, an enantiomerically pure compound has been required in stoichiometric amounts, although in some cases it could be recovered for reuse. The final refinement, possible in the fourth-generation methods, is to use a *chiral catalyst* to direct the conversion of an achiral substrate directly to a chiral product with an achiral reagent. Again the control here is intermolecular:

$$S \xrightarrow[\text{cat.}^*]{R} P^*$$

An example is the asymmetric conjugate addition of a thiophenol to cyclohexenone to give (**42**) catalysed by the alkaloid cinchonidine.

This class of reaction, which includes many enzyme-catalysed transformations, is obviously very attractive and is the subject of intensive research at the present time. By definition the catalyst can be recovered unchanged at the end of the reaction and in many cases only a small quantity (0.05 equiv. or less) is required, although this is not always the case and a larger quantity may have to be used, either to achieve a reasonable rate of conversion or to account for losses due to side reactions. The possibility of using only a 'catalytic amount' of enantiomerically pure compound obviously has great attractions for large-scale industrial use where cost is critical. A detailed discussion of third- and fourth-generation methods is the subject of chapter 6.

4.2.4 Special methods

Kinetic resolution[6]

This procedure involves reaction of either a racemic chiral compound or an achiral compound containing equivalent enantiotopic groups with a chiral reagent or an achiral reagent/chiral catalyst system. In either case the two enantiomers or enantiotopic groups undergo reaction at different rates and in the ideal case one enantiomer (enantiotopic group) is converted to product while the other remains unchanged i.e.:

$$(\pm)\text{-S} \xrightarrow[\text{or } \mathbf{R/cat}^*]{\mathbf{R}^*} (+)\text{-S}^* + (-)\text{-P}^*$$

Several features of this type of reaction are notable. In the simple case shown, as in a classical resolution, the maximum yield of one product is 50% and the e.e. actually varies as the reaction progresses due to the kinetics of the system. If, however, the reaction is carried out under conditions in which the enantiomers of the substrate can interconvert (racemisation), the entire substrate can in principle be converted to the enantiomerically pure product and the e.e. of the product then remains constant throughout the course of the reaction. A good illustration of this point is provided in section 6.5.1.

Many of the most useful applications of enzymes in asymmetric synthesis involve kinetic resolution[7] and an example is the hydrolysis of (±)-N-acetylphenylalanine methyl ester (**43**) with α-chymotrypsin to give the (S)-acid (**44**) and the unchanged (R)-ester (**45**). Very often, as in this case, we can make use of either of the two products once they have been resolved by a further simple non-asymmetric chemical step (here hydrolysis of (**45**) to give the (R)-acid).

The other important type of kinetic resolution is that in which the chiral reagent or catalyst discriminates between two enantiotopic groups in an achiral substrate. This may be thought of as a kinetic resolution within the same molecule and the substrate can be completely converted to a single enantiomeric product. Thus diester (**46**) is hydrolysed by pig liver esterase (PLE) to give exclusively the (S)-enantiomer of (**47**). A drawback of the 'internal' kinetic resolution is that it may not be possible to find a catalyst to obtain the opposite enantiomer of the product if this is wanted. This has been possible in some cases as illustrated by the conversion of the *meso* diacetate (**48**) to the diol monoacetate (**49**) using PLE but to the opposite enantiomer (**50**) using the lipase from *Candida cylindracea*.

Double and triple asymmetric induction

An important development in recent years has been the introduction of more sophisticated methods which, on the face of it, combine elements of the first-, second-, third- and fourth-generation methods described in the last section. Thus we can, for example, react a chiral substrate with a chiral reagent. This approach, pioneered by Masamune,[8] is described as multiple asymmetric induction and is particularly valuable in reactions where two new stereogenic units are formed simultaneously. In the example which follows, the aldol reaction between chiral aldehyde (**51**) and a boron enolate to give (**53**) and (**54**) is considered. As it stands this is a first-generation method since the chiral substrate (**51**) reacts with the achiral enolate to give

the product with new stereogenic centres. Even with the achiral enolate (**52**), the *relative* stereochemistry of the product is controlled by Cram's rule and only the two isomers shown are formed, having the *u*-configuration at the new centres. This would also be the case for an achiral aldehyde component, but the chirality of (**51**) obviously has a small effect in controlling the absolute stereochemistry since the ratio of (**53**) to (**54**) formed using (**52**) is 3:2 rather than 1:1. If we now replace (**52**) by the chiral boron enolate reagent (**55**), double asymmetric induction occurs and the same two products are formed but the ratio is now >100:1 in favour of (**53**). It is most interesting, and in fact characteristic of multiple asymmetric induction processes, that the selectivity obtained by using the opposite enantiomer of one component is generally opposite but not equal. Thus reaction of (**51**) with the enantiomeric enolate (**56**) gives a product ratio of only 30:1 in favour of (**54**). This observation is explained in terms of 'matched' and 'mismatched' pairs of reactants as described in more detail in section 5.3.3.

More recently the first examples of triple asymmetric induction have been described. Keeping with the example of the asymmetric aldol reaction, this would involve, for example, reaction of chiral substrate (**51**) with a boron enolate which has not only a chiral R group as in (**55**) but also a chiral auxiliary attached in the form of the BR^1_2 group, i.e. (**57**). Not surprisingly, this allows for further possible combinations of matched and mismatched components and the picture becomes quite complex. However, it is clear that this approach allows very high selectivity in the formation of several stereogenic centres and will become increasingly important.

4.3 Mechanistic considerations

Virtually all diastereoselective and enantioselective reactions are based on a kinetic phenomenon: if the rate constant k_R of the reaction leading to the R product is greater than k_S leading to the S product, then the product with configuration R at the new stereogenic centre will predominate, and vice versa.

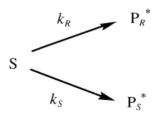

The enantiomeric or diastereomeric ratio is simply the ratio of the rate constants.

$$\frac{[P_R^*]}{[P_S^*]} = \frac{k_R}{k_S}$$

The Arrhenius equation gives the relationship between the rate constant and the activation energy.

$$k_R = Ae^{-E_R/RT}, \quad k_S = Ae^{-E_S/RT}$$

$$\therefore \quad \frac{[P_R^*]}{[P_S^*]} = e^{-(E_R - E_S)/RT} = e^{-\Delta E/RT}$$

In other words, the greater the difference in the activation energy of the two pathways, the greater the selectivity. When the activation energy of the reaction giving the *R* product is the same as that of the reaction to the *S* product, there is no selectivity, and the ratio is 1:1. The difference in activation energy need not be very large to obtain high diastereo- or enantioselectivity: at 300 K a difference of 2 k cal mol-1 (8.4 kJ mol^{-1}) gives a ratio of 96:4.

It is important to note the dependence of the selectivity achieved on the reaction temperature. For a given value of ΔE, lowering the temperature will lead to increased selectivity (but also to a slower rate). Many asymmetric reactions are carried out well below room temperature, in some cases as low as $-120°C$, in order to achieve the maximum selectivity.

Let us now consider in more detail the nature of this difference in activation energy. The transition states leading to the *R* and *S* products are *diastereomeric*, and therefore *non*-equivalent, and of different energies. In most of the stereodifferentiating reactions described in this book the key step involves preferential addition from the *Re* or *Si* face to a trigonal carbon, which is usually unsaturated.

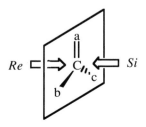

For the reaction to be stereodifferentiating, these diastereotopic faces must be non-equivalent. This non-equivalence may arise in three basic ways. First, there may be an adjacent stereogenic centre in the substrate which, for example, renders one face more sterically encumbered than the other. This is the basis of first- and second-generation methods. Second, the attacking species may contain a stereogenic centre which causes it to interact more favourably with one face than the other. This is, of course, the case in third-generation, reagent-controlled asymmetric reactions. Third, neither the substrate nor the attacking species on their own may be chiral but the attack is mediated by a chiral catalyst which renders the faces non-equivalent. This is the fourth-generation, chiral catalyst method. At the heart of asymmetric synthesis there lies this non-equivalence of the diastereotopic faces, with the consequent difference in the activation energies. The art of the organic chemist is in maximising the difference.

References and notes

1. J.W. Scott, in *Asymmetric Synthesis*, Vol. 4, J.D. Morrison and J.W. Scott, eds, Academic Press, 1984, Chapter 1.
2. A recent monograph describes asymmetric syntheses starting from amino acids, grouped according to the starting material used: G.M. Coppola and H.F. Schuster *Asymmetric Synthesis—Construction of Chiral Molecules Using Amino Acids*, Wiley, 1987.
3. For a recent review see H. Wynberg, Asymmetric catalysis by alkaloids, *Top. Stereochem.*, 1986, **16**, 87.
4. This approach is described in detail in the monograph by S. Hanessian, *Total Synthesis of Natural Products: The Chiron Approach*, Pergamon, Oxford, 1983. See also T.D. Inch, *Tetrahedron*, 1984, **40**, 3161.
5. P. Newman, *Optical Resolution Procedures for Chemical Compounds*, Vol. 1–3, Optical Resolution Information Centre, New York, 1981–84.
6. H.B. Kagan and J.C. Fiaud, *Top. Stereochem.*, 1988, **18**, 249.
7. See H.G. Davies, R.H. Green, D.R. Kelly and S.M. Roberts, *Biotransformations in Preparative Organic Chemistry*, Academic Press, 1989.
8. S. Masamune, W. Choy, J.S. Petersen and L.R. Sita, *Angew. Chem., Int. Ed. Engl.*, 1985, **24**, .

5 First- and second-generation methods: chiral starting materials and auxiliaries

S.N. KILENYI

We have defined first- and second-generation asymmetric synthesis as follows: a first-generation method starts with an enantiomerically pure compound (almost invariably of natural origin) which is incorporated into the final product, whereas in a second-generation method a chiral auxiliary is attached and, after serving its purpose, is then removed. In each case asymmetry in the newly formed stereogenic centres is induced intermolecularly by those already present. Though perhaps not as 'elegant' as third- and fourth-generation asymmetric syntheses, these methods have the advantage that, providing the chiral starting material or auxiliary is enantiomerically pure, the final product may be obtained with very high e.e., since any loss of enantioselectivity gives rise to separable *diastereomers* rather than enantiomers. On the other hand, there is the disadvantage that two extra steps are necessary in second-generation asymmetric syntheses: to introduce, then remove, the chiral auxiliary.

The bulk of this chapter is concerned with second-generation methods, subdivided according to whether the reactant bearing the chiral auxiliary reacts as a nucleophile, an electrophile, or neither. We begin, however, by considering a few examples of non-stereodifferentiating and first-generation methods.

5.1 Non-stereodifferentiating methods

5.1.1 Classical resolution ((1R)-cis-permethric acid)

Although the classical method of obtaining enantiomerically enriched chiral compounds by resolution is not within our brief, an example is given here for the sake of completeness. The work cypermethrin

(1R)-cis-(2)

illustrates the point that, with recycling of the undesired enantiomer, resolution can be a valuable method, especially on a plant scale. In the example opposite, the target is (1R)-*cis*-permethric acid (**2**), which is part of the important insecticide cypermethrin.

The racemic mixture of cyclobutanones is converted into bisulphite adducts (note that diastereoselective attack of the bisulphite anion takes place exclusively *anti* to the two adjacent substituents). Fractional crystallisation of the salt formed with enantiomerically pure (*S*)-1-phenylethylamine allows separation of the (2*S*, 4*R*)-diastereomer (**1**), which, after decomposition of the bisulphite adduct, undergoes a Favorskii-type ring contraction and elimination of HCl to give enantiomerically pure (**2**). The undesired (2*R*, 4*S*) diastereomer can be converted back to the racemic mixture for recycling, rendering the process highly efficient.[1]

5.1.2 Resolution using a chiral auxiliary

Resolution of enantiomers may also be performed by the *covalent* attachment of an enantiomerically pure compound more comonly used as a chiral auxiliary, followed by separation of the diastereomers. A popular choice for ketones and lactones is (+)-(*R*,*R*)-2,3-butanediol, whose use is illustrated here by the synthesis of a fragment of (+)-latrunculin B.[2]

Alkylation of the racemic lactone (**3**) gives the diastereomers in a 1:1 ratio and consists of an equal mixture of both enantiomers. The chiral diol is installed to give the orthoester of the lactone, which results in the equilibration of the *syn* isomer in a ratio of 6:1. At this point the diastereomers of the major *anti* isomer can be separated by HPLC, which also serves to remove the minor *syn* products. Ozonolysis of (**4**) gives the aldehyde, whose two stereogenic centres become C-8 and C-11 of (+)- latrunculin B. The *absolute* configuration of these centres could not be predicted in advance, so a crystalline derivative was made, allowing an X-ray determination of the structure. Since the absolute configuration of the (*R*, *R*)-2,3-butanediol is known with certainty, that of the orthoester could be inferred with confidence.

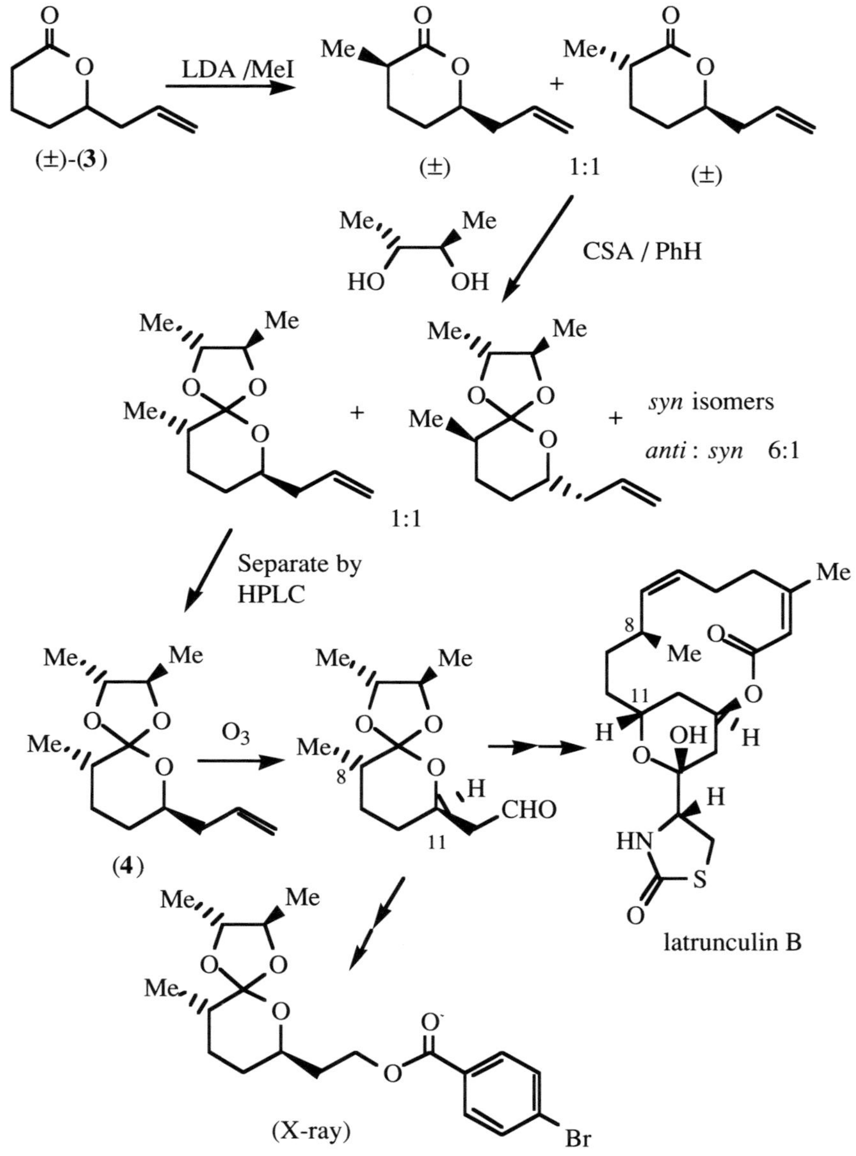

ASYMMETRIC SYNTHESIS

latrunculin B

(4)

(X-ray)

5.1.3 Sulphoximines

We saw in section 1.8 that while the two oxygens of an unsymmetrical sulphone are enantiotopic, if one of the oxygens is replaced by a nitrogen, the resulting compound is known as a sulphoximine, and is chiral. Johnson has developed one such compound as a combination chiral auxiliary/resolving agent.[3] The reagent, (S)-(+)-N,S-dimethyl-S-phenyl sulphoximine (5), is synthesised from racemic methyl phenyl sulphoxide and resolved with (+)-10-camphorsulphonic acid.

The S-methyl group of (5) can be deprotonated with strong base. The resulting anion (6) undergoes nucleophilic addition to prochiral ketones, generating the two diastereomers (7) and (8) in unequal quantities. Separation of these on a silica column followed by Raney nickel desulphurisation gives both tertiary alcohols (9) and (10) with variable e.e.

In fact this hardly qualifies as a stereodifferentiating reaction, since the diastereofacial selectivity on the prochiral ketone is quite low and unpredictable. Nonetheless the method is useful, especially with aryl ketones which allow claen separation of the diastereomers, giving e.e.s in the range 85–95%. For aldehydes and dialkyl ketones the results are poorer.

Secondary alcohols (**11**) and (**12**) may also be made with good enantiomeric purity by a variant of the method.

FIRST- AND SECOND-GENERATION METHODS

The most interesting application of chiral sulphoximine technology lies in asymmetric cyclopropanation.[3] This relies on the fact that the Simmons–Smith reagent can be directed exclusively to one face of a double bond by an adjacent hydroxyl group through complexation. Here, the sulphoximine serves in its usual role of chiral 'handle' to render the process asymmetric.

The final step illustrates the fact that the chiral auxiliary can be removed for reuse by mild thermolysis to give the products (**13**) and (**14**).

A further elegant application of the chiral sulphoximine (**5**) as a combined resolving agent and chiral auxiliary is to be found in the synthesis of chiral (−)-thujopsene.[4]

5.1.4 Methods using enantiomerically pure building blocks

We conclude this section with two examples of the synthesis of chiral products by starting from an appropriate chiral building block. This has been named the 'chiron' approach by Hanessian, and is described in great detail in his book.[5] The distinctive feature of this type of reaction is that no new stereogenic units are formed and all the stereogenic units of the product are directly derived from the starting material. For this reason these cannot be considered as asymmetric synthesis but methods of this type are of some practical importance in areas where suitable chiral starting materials are readily available.

The first application involves an amino acid as chiral starting material in the synthesis of *un*natural amino acids, those never imagined by nature.[6] *S*-Serine (**15**) is converted into its N-protected analogue

FIRST- AND SECOND-GENERATION METHODS

(**16**). Under Mitsunobu conditions the primary hydroxyl is displaced, giving (**17**). This strained β-lactone undergoes $S_N 2$ ring-opening rather than the normal attack on the carbonyl when treated with a wide variety of Grignard reagents in the presence of Cu(I) salts. To complete the synthesis of the unnatural amino acid, the CBz group is removed to give the enantiomerically pure product (**18**). The opposite enantiomer can be made from the commercially available *R*-serine.

The second example in this section is taken from some work performed by chemists at Squibb.[7] The target molecule is aztreonam, which is a monobactam antibiotic currently in clinical trials.

Natural (2*S*, 3*R*)-threonine is converted into hydroxamate (**19**) by standard means. An internal $S_N 2$ displacement of mesylate occurs on base treatment, giving β-lactam (**20**). Note the inversion of the stereogenic centre bearing the mesyloxy group. The *N*-methoxy group, having served its function of acidifying the NH, is removed reductively, and the β-lactam sulphated to give (**21**). The synthesis is completed by deprotection of the primary amine and coupling with the rest of the molecule. Overall, all the carbons of threonine are used, with one stereocentre retained and the other inverted, which is a highly efficient use of this amino acid.

5.2 First-generation methods

As described in chapter 4, virtually all the known methods of asymmetric synthesis are ultimately based on the pool of naturally occurring enantiomerically pure compounds such as amino acids, terpenes and sugars produced by living organisms. For example, the source of molecular asymmetry in the Sharpless epoxidation is tartaric acid obtained, as in Pasteur's day, by fermentation with yeast.

In this section we describe a few representative examples of first-generation asymmetric syntheses, in which part or all of the starting chiral compound is actually built into the final product and serves to direct the formation of the new stereogenic centres. The syntheses below are divided into three groups on the basis of the chiral starting material used.

5.2.1 Sugars

The plant kingdom provides enantiomerically pure sugars in abundance which can be converted by relatively standard methods into more interesting compounds. Avenaciolide is a naturally occurring α-methylene lactone with cytotoxic properties. Fraser-Reid and Anderson synthesised it from glucose,[8] having recognised the high degree of overlap between the sugar and the target molecule. All but one of the carbons and two of the five stereogenic centres of glucose are used. As the reader will recall, glucose is in equilibrium with a small amount of the furanose form, which reacts with 2,2-dimethoxypropane to give the (commercially available) diisopropylidene-α-D-glucofuranose (**22**). Oxidation of the remaining hydroxyl and Horner-Emmons reaction of the resulting ketone gives (**23**). In the asymmetric step this is reduced stereoselectively to (**24**), which now contains all the stereogenic centres of the target molecule in their correct absolute configuration. The rest of the synthesis is unexceptional: selective cleavage of the pendant ketal group (well precedented in sugar chemistry), cleavage of the vicinal diol to (**25**), homologation of the aldehyde to (**26**), hydrolysis of the remaining ketal with concomitant lactonisation giving (**27**), and finally oxidation of the lactol and installation of the exo-methylene group to give avenaciolide which is essentially enantiomerically pure.

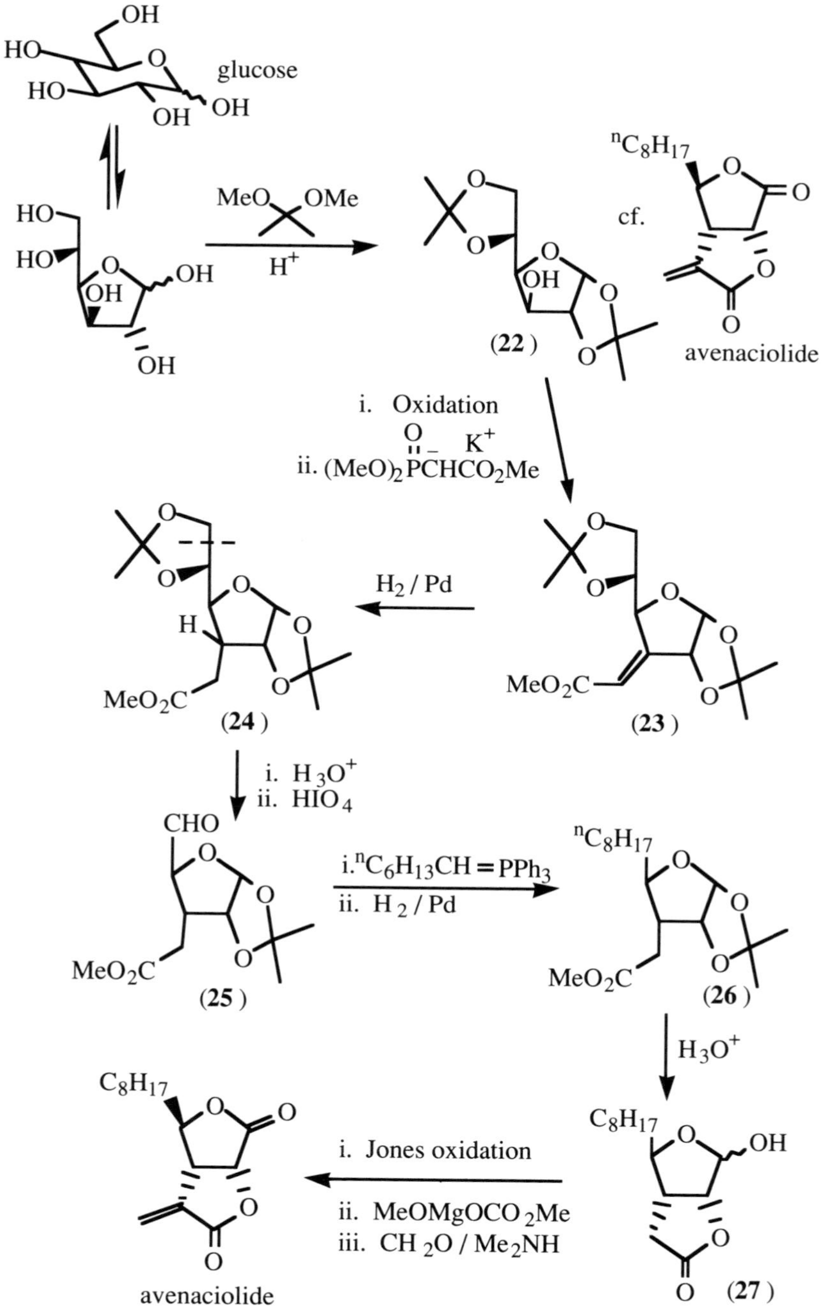
OH
HO
O
glucose
HO
OH
OH
HO
HO
O
OH
OH
OH
MeO
OMe
H+
O
O
O
OH
O
O
(22)
cf.
nC8H17
O
O
O
O
avenaciolide
i. Oxidation
ii. (MeO)2PCHCO2Me
K+
O
O
O
O
H
O
O
MeO2C
(24)
H2 / Pd
O
O
O
O
O
MeO2C
(23)
i. H3O+
ii. HIO4
CHO
O
O
O
MeO2C
(25)
i. nC6H13CH=PPh3
ii. H2 / Pd
nC8H17
O
O
O
MeO2C
(26)
H3O+
C8H17
O
O
O
O
avenaciolide
i. Jones oxidation
ii. MeOMgOCO2Me
iii. CH2O / Me2NH
C8H17
O
OH
O
O
(27)

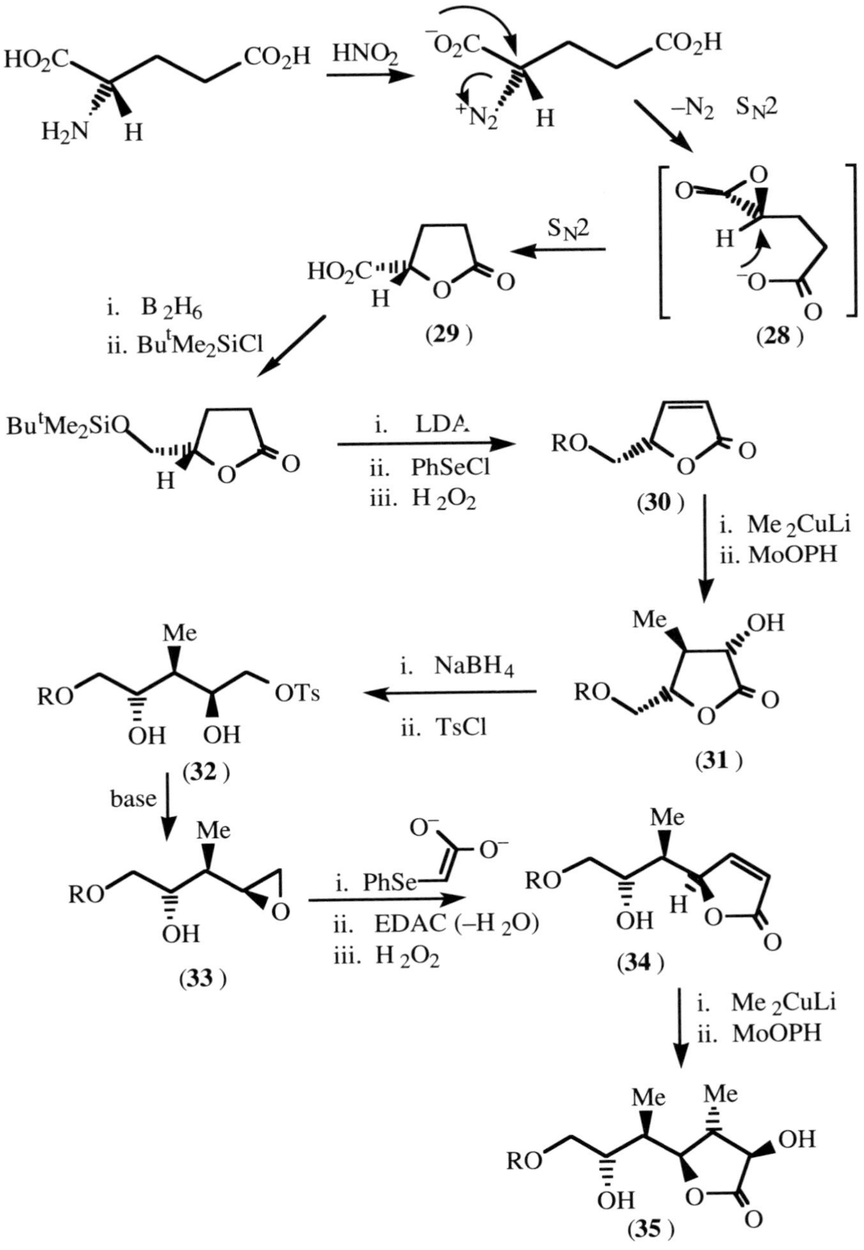
HNO2
–N2 SN2
SN2
(29)
(28)
i. B2H6
ii. ButMe2SiCl
i. LDA
ii. PhSeCl
iii. H2O2
(30)
i. Me2CuLi
ii. MoOPH
i. NaBH4
ii. TsCl
(31)
(32)
base
(33)
i. PhSe
ii. EDAC (–H2O)
iii. H2O2
(34)
i. Me2CuLi
ii. MoOPH
(35)

5.2.2 Amino acids

This example, the work of Hanessian, shows how asymmetry may be 'grown' along a carbon chain by the use of 1,2-diastereoselection.[5,9] The starting material is (S)-glutamic acid (whose monosodium salt is the famous flavour enhancer) and the target is a typical polypropionate-derived chain found in many macrolide antibiotics.

Diazotisation of (S)-glutamic acid results in loss of N_2 with formation of strained α-lactone (**28**) via an internal S_N2 displacement, followed by a second S_N2, giving α-lactone (**29**) with overall *retention*. Conversion to unsaturated lactone (butenolide) (**30**) is standard. Lithium dimethylcuprate adds with 1,2-diastereoselection exclusively *anti* to the silyloxymethyl substituent, and the resulting enolate undergoes electrophilic hydroxylation *anti* to the newly formed stereocentre giving (**31**).

Conversion of (**32**) into (**34**) via epoxide (**33**) needs no comment. The same trick can then be applied to (**34**) giving (**35**), which has no fewer than five stereogenic centres, four of them installed with almost total fidelity under the direction of the original centre from glutamic acid. Thus, the butenolide is used as a chiral *template* to propagate the asymmetry of glutamic acid along the chain. By variations on the theme of the conversion, virtually all imaginable configurations of a polypropionate chain can be synthesised enantiospecifically.

5.2.3 Terpenoids

Terpenes and terpenoids are popular chiral starting materials because they are cheap, available in bulk and chemically versatile.

So far, the synthetic sequences have been linear. It should be clear that in a convergent synthesis, the two or more fragments to be joined *must* be enantiomerically pure to avoid impossible difficulties with diastereomeric mixtures. This point is well illustrated by the Williams synthesis of the natural enantiomer of the complex antibiotic milbemycin β_3,[10] in which the ultimate sources of chirality are the odiferous terpenoid (−)-(3S)-citronellol and the carbohydrate (+)-mannitol.

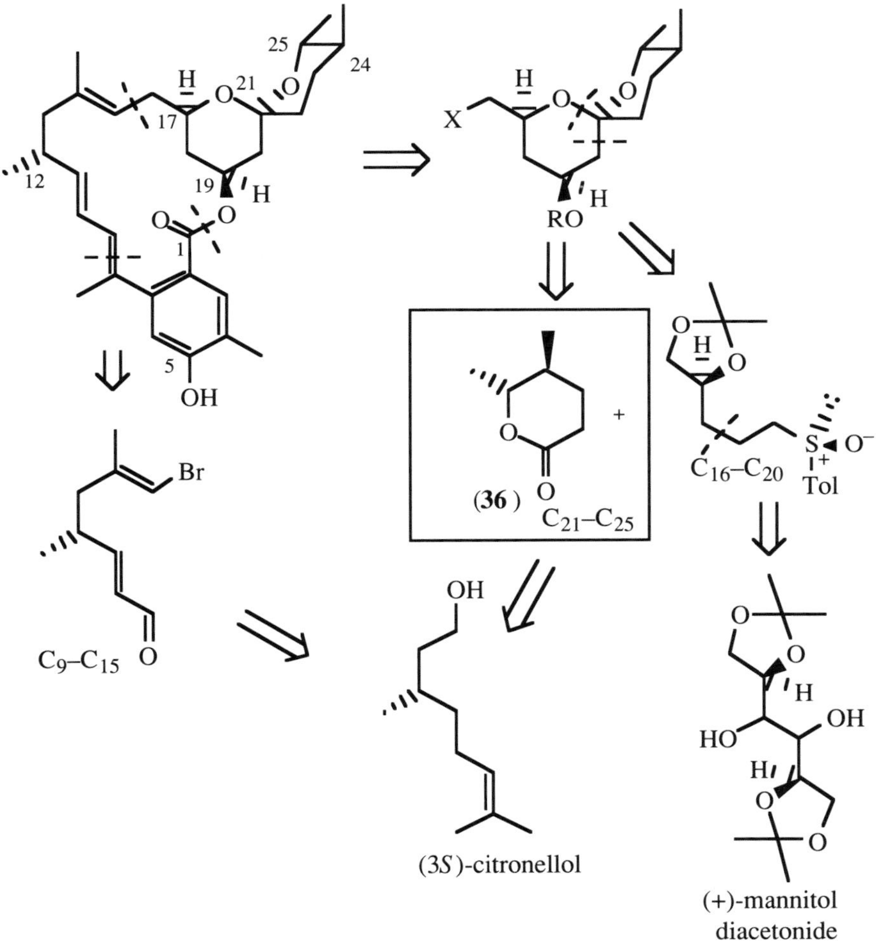

(3S)-citronellol

(+)-mannitol diacetonide

We will only discuss the synthesis of lactone (**36**), which constitutes the C_{21}–C_{25} section of milbemycin β_3. (3S)-Citronellol is dehydrated to (**37**), which is selectively ozonised at the more electron-rich double bond, giving aldehyde (**38**). Jones oxidation of (**38**) furnishes acid (**39**). The key reaction is the iodolactonisation of (**39**). The molecule is believed to adopt a chain-like conformation with the methyl group in an equatorial position. Ring closure to (**40**) occurs with very high (15:1) diastereomeric, and therefore enantiomeric, excess. Finally, the iodide is reduced under radical conditions to give the C_{21}–C_{25} lactone (**36**).

See also section 7.1.1 for another application of terpenoids as chiral starting materials.

5.2.4 Hydroxy acids

As an example of a straightforward asymmetric synthesis based on a hydroxy acid, we have chosen that of HR 780, an inhibitor of the

FIRST- AND SECOND-GENERATION METHODS 99

enzyme HMG CoA reductase, which Hoechst hopes to commercialise for the treatment of atherosclerosis.[11] The synthesis begins with (S)-malic acid, which is transformed by unexceptional chemistry into (**41**). The ester function is homologated by a directed Claisen condensation to hydroxy ketoester (**42**). The second stereogenic centre is then installed by a stereospecific reduction of the ketone to give (**43**). The reason for the selectivity is that the reductant is delivered *intramolecularly* by the neighbouring hydroxyl, leading to excellent 1,3-diastereoselection. The rest of the synthesis is relatively standard.

We will see many other examples of this important tactic of controlling diastereoselectivity by intramolecular delivery. For another asymmetric synthesis of the lactone of mevinolin, see section 7.4.5.

The second example demonstrates that there need not be an obvious overlap between the structures of the chiral starting material and the target molecule, in this case the β-lactam antibiotic (+)-PS-5.[12] The chiral starting material (S)-lactic acid is transformed by standard means into (S)-2-hydroxypropanal (**44**), which is converted into its O,N-disilylimine (**45**). The crucial step involves the reaction of (**45**) with the Z-enolate of t-butylbutanoate.

Two new stereogenic centres are formed with very high *anti* diastereoselectivity with respect to each other, and high diastereoselectivity with respect to the existing asymmetric centre. Since the starting imine is enantiomerically pure, the two new stereogenic centres are also formed with high *enantio*selectivity. We will consider first the role of the asymmetric centre in the imine. The selectivity at the imine derived centre is the result of the so-called 'internally-chelated' Cram's rule. The oxygen and nitrogen chelate the electropositive lithium cation, forming a cyclic species (**48**). The two faces of the imine are non-equivalent, the *ul (Re, S)*

being hindered by the methyl group. Nucleophilic attack therefore occurs preferentially from the *lk* face. The authors claim that the 1,2-*anti* selectivity at the ester derived centre derives from a boat-shaped transition state which is held together by multiple chelation.

The rest of the synthesis is almost an anticlimax. The β-lactam (**46**) is formed spontaneously and then desilylated to (**47**). Standard reactions are then employed to convert (**47**) into the target compound. Note that the original stereogenic centre is lost!

5.3 Second-generation methods: nucleophiles bearing a chiral auxiliary

5.3.1 General principles

As already described in section 4.2.3, the use of a chiral auxiliary in asymmetric synthesis comprises three steps: (i) installation of the enantiomerically pure auxiliary on the substrate; (ii) reaction with an achiral reagent producing the two possible diastereomers in unequal quantity; (iii) removal of the auxiliary *without* racemisation.

The transition states leading to the R and S products are *diastereomeric* and therefore different in energy. We have seen in section 4.3 how the diastereomeric ratio is related to the two rate constants k_R and k_S and the activation energies E_R and E_S for formation of the diastereomeric transition states. The art of designing chiral auxiliaries lies in maximising this difference in activation energies. What is the origin of this difference?

The factors directing attack on one face or the other are usually of the following type: first and foremost steric, but also chelation of metal cations, hydrogen bonding and electrostatic interactions. Generally, for high diastereofacial selectivity, a rigid transition state with many contacts between the partners is necessary.

The second-generation methods described in this chapter have been divided into three main groups. In the first, which take up the rest of this section, the substrate bearing the chiral auxiliary reacts as a nucleophile, in the second (section 5.4) it is electrophilic, and in the third (section 5.5) it is neither.

5.3.2 Chiral enolate and aza-enolate equivalents

The enolate, perhaps the most versatile synthetic intermediate of all, lends itself well to asymmetric synthesis via chiral auxiliaries. There are three principal sites for the installation of an auxiliary, A^*: either on C-1 of the enolate or on the nitrogen atom of the 'aza-enolates' derived from imines or amides:

All these strategies have been explored with great success.

The hydrazine SAMP, derived from (*S*)-proline, and its enantiomer RAMP were developed by Enders[13] for the asymmetric alkylation of aldehydes and ketones, and are commercially available, though expensive. The chiral auxiliary may be removed either by quaternisation with methyl iodide followed by hydrolysis, or by ozonolysis.

In the first example, 3-pentanone is converted via its SAMP hydrazone (**49**) into an ant alarm pheromone (**50**) with an outstanding enantiomeric purity.

The explanation of the very high diastereofacial selection is not yet fully understood, though it is quite clear that the hydrazone anion has a rigid, internally chelated structure involving the lithium cation and the methoxy oxygen, making the *Re* and *Si* faces highly non-equivalent. The observed selectivity is consistent with the electrophile attacking from the apparently more hindered *ul (Re, S)* face and it may well be 'guided in' by complexation of I to Li$^+$. Reaction of (**49**) with a more complex electrophile to give another natural product of insect origin is described in section 7.2.6.

The SAMP and RAMP hydrazone anions also undergo Michael additions with excellent asymmetric induction at the electrophilic carbon. (Note that the configuration of the imine is unimportant—the LDA deprotonates exclusively at the methyl group.)

The chiral auxiliary may be recycled without difficulty simply by mild reduction of the *N*-nitroso group.

Probably the most versatile chiral auxiliary of all is the oxazoline of Meyers.[14] This auxiliary provides a simple chiral equivalent of an ester enolate. The oxazoline (**52**) is formed as shown from the (*S,S*)-aminodiol (**51**), which is a byproduct from a microbial step in the production of chloramphenicol and is commercially available. The observed selectivity depends on exclusive formation of the Z-azaenolate (**53**) followed by attack of the electrophile from below due to its complexation with the lithium.

This reaction is extremely versatile and a wide variety of different electrophiles have been used. We will see other examples of chiral oxazoline chemistry later in section 5.4.1.

Meyers has published a particularly simple and efficient asymmetric synthesis of cyclohexenones (**54**) bearing a quaternary stereogenic centre at position 4 which again uses aminodiol (**51**) as the source of chirality.[15]

The value of the method lies in the fact that quaternary centres are fairly difficult to make asymmetrically by other methods. Once again, the diastereofacial selectivity derives from the cyclic chelate structure of the enolate alkoxide, though the exact nature of the species is as yet unclear. A useful feature is that the opposite enantiomer of the product can readily be obtained by simply reversing the order of the two alkylation steps.

The cyclic diamides formed by the cyclodehydration of α-amino acids are known trivially as 2,5-diketopiperazines, and are the important element of an asymmetric synthesis of α-methyl substituted amino acids developed by Schöllkopf.[16] The anion (**55**) resulting from deprotonation of the bis-lactim ether of a diketopiperazine undergoes alkylation with very high diastereofacial selection, especially if the chiral auxiliary is derived from (*S*)-valine.

Thus the overall result is C-alkylation of (S)-alanine with inversion. The method also works well in the asymmetric alkylation of glycine, as shown by the asymmetric synthesis of (S)-phosphonothricin (**56**), a naturally occurring herbicide.[17] Note that the configuration at the tetrahedral phosphorus is irrelevant, since proton exchange in the phosphinic acid racemises this centre.

108　　　　　　　　　ASYMMETRIC SYNTHESIS

Because of their occurrence in many biologically active compounds, chiral α-amino acids are among the most important targets for asymmetric synthesis. A wide variety of methods have been developed, in addition to the two included in this section, and these are described in a recent monograph.[18]

The Evans oxazolidinone methodology is quite versatile and quite apart from its use in aldol reactions (section 5.3.3) lends itself well to the asymmetric synthesis of carboxylic acids substituted in the α-position with oxygen, nitrogen, and carbon.[19] The auxiliary shown is derived from (+)-norephedrine and the opposite enantiomers of the products are available from the valine-derived auxiliary.

The final example of asymmetric enolates concerns the iron acyl complexes developed by the Davies group.[20] These complexes give excellent enantioselectivity in a wide variety of asymmetric reactions but unfortunately the auxiliary can only be obtained by a tedious resolution procedure, is very expensive and cannot readily be recycled.

As in all the previous examples, the two faces of the enolate are non-equivalent. In the complex shown, the *Re* face is heavily encumbered by a phenyl group, leading to virtually complete diastereoselection from the *Si* face. The product is the active enantiomer of captopril, an important antihypertensive drug.

5.3.3 Asymmetric aldol reactions[21]

In its most general form the aldol reaction can be represented by the general formula shown.

There are three questions to be addressed: first, the relative stereochemistry of the new stereogenic centres C-2 and C-3 with respect to each other; second, the influence of asymmetry in the aldehyde A on C-2 and C-3; third, the influence of asymmetry in the enolate B.

Let us consider first the 2,3-stereochemistry. There are two new stereogenic centres, giving rise to four possible outcomes:

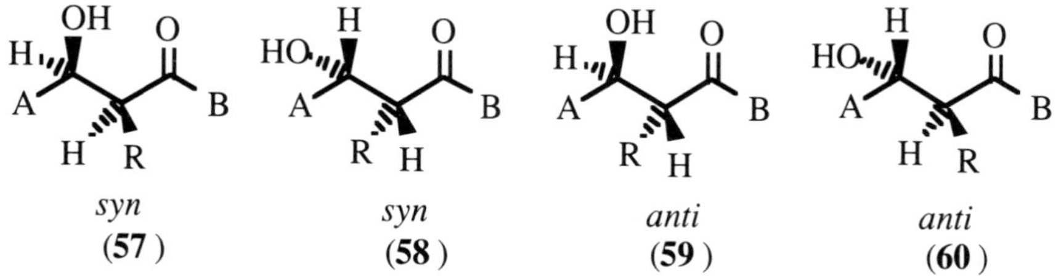

syn
(**57**)

syn
(**58**)

anti
(**59**)

anti
(**60**)

The two *syn* aldols (**57**) and (**58**) are enantiomers (provided there is no additional asymmetry in A or B), as are the *anti* aldols (**59**) and (**60**). The *syn*/*anti* outcome depends fundamentally on the geometry of the enolate and can be predicted on the basis of the six-membered cyclic transition state known as the Zimmermann–Traxler model.

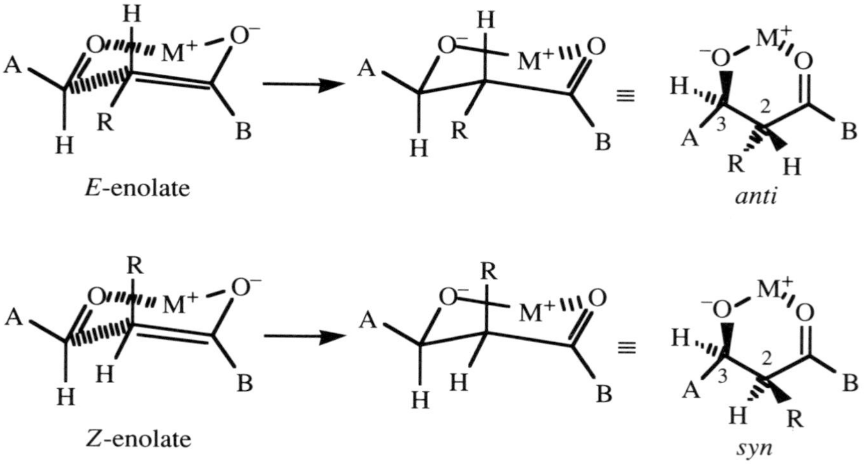

E-enolate

anti

Z-enolate

syn

FIRST- AND SECOND-GENERATION METHODS

Thus, assuming that the *oxygen* atom of the enolate takes precedence over group B, we can see that the *E*-enolate gives rise to the *anti*-aldol, and the *Z*-enolate to the *syn*. Note the resemblance to the chair conformer of cyclohexane, in which the bulky A group occupies the equatorial position. In practice, the best *syn/anti* selectivity is obtained with M = Li, Mg or BR_2. Thus the problem of relative 2,3-diastereoselection reduces to the preparation of the necessary *E*- or *Z*-enolate, which is not always simple.

Next, we must consider the influence of a stereogenic centre adjacent to the aldehyde carbonyl, whose faces now become enantiotopic. The diastereoselection here is predicted by the Cram model. The three ligands on the stereogenic centre are supposed to be of different steric bulk: large (L), medium (M) and small (S). In the Cram model the aldehyde (**61**) is supposed to prefer the conformation (**62**) in which the carbonyl bisects the small–medium sector, to minimise steric interaction.

Attack of the nucleophile then occurs preferentially from the less-hindered side, the *Re* in this case, to give product (**63**). Unfortunately, the 1,2-diastereoselection obtained is rarely better than about 5:1, and many exceptions are known (so-called 'anti-Cram' products).

If one of the groups on the stereogenic centre is capable of chelating a metal cation, a different diastereoselectivity is observed due to the formation of a cyclic chelate.

This time the 1,2-diastereoselectivity can be very high because of the rigidity of the chelate.

Now let us consider the reaction between an achiral enolate and a chiral aldehyde. As we have seen, the 2,3-diastereoselection is controlled only by the enolate geometry. However, the diastereofacial selection, and therefore the relation between the new centres and that on the aldehyde, is determined by Cram's rule or its variant.

An example should make this clear. The aldehyde (**64**) carries a chelating group, suggesting that, in the presence of magnesium bromide, the facial selectivity is of the 'chelated' Cram-rule type.[22] Note the *syn* selectivity due to the Z-enolate (**65**).

The diastereofacial selectivity of an asymmetric aldol reaction can also be controlled on the enolate side, and this is the basis of the second-generation methods of Evans[23] and Masamune.[24] The complementary Evans auxiliaries (**66**) and (**67**) are synthesised from (*S*)-valine and (1*S*,2*R*)-norephedrine respectively. The *Z*-enolate (**68**) is formed exclusively on reaction with dibutylboron triflate, and this reacts with aldehydes to give essentially only one aldol product (**69**). The diastereofacial selectivity derives from the bulky groups on the auxiliaries which force attack from the opposite face.

As can be seen from the diagram, the disfavoured transition state suffers a severe clash between (in this case) the isopropyl group and the enolate portion. The opposite enantiomeric series (**70**) is available from the norephedrine-derived auxiliary (**67**).

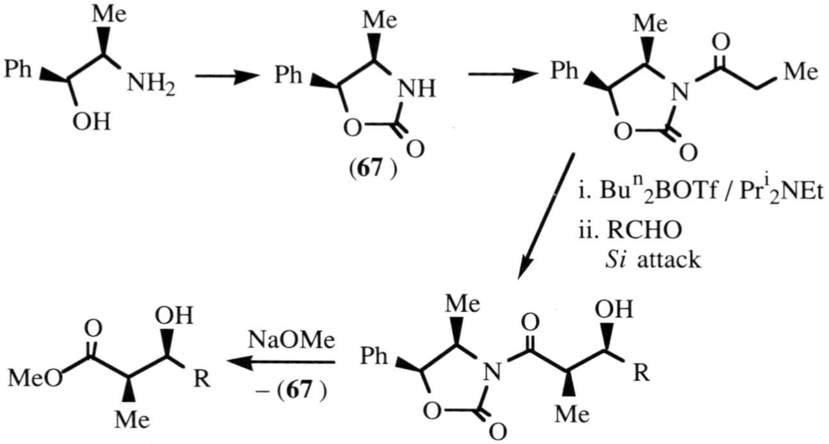
Me
Ph
NH2
OH
NH
O
(67)
N
i. Bu^n_2BOTf / Pr^i_2NEt
ii. RCHO
Si attack
NaOMe
– (67)
MeO
R
OH
(70)

If the counter-ion is lithium or sodium, the topicity is reversed because of the cyclic chelate formed.

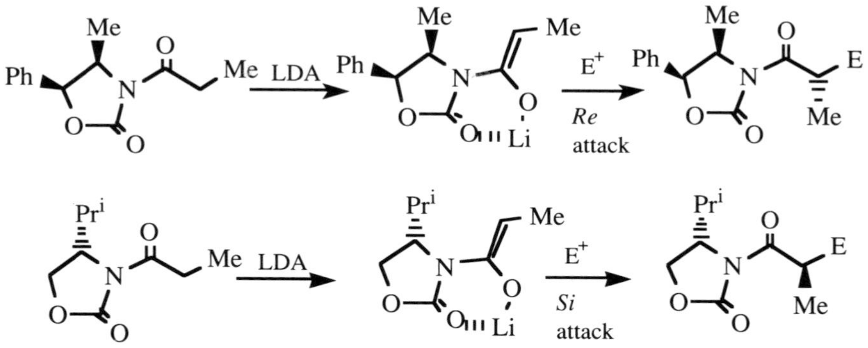
Me
Ph
N
O
LDA
Li
E+
Re
attack
E
Pr^i
Si
attack

Further examples of Evans' asymmetric aldol method are seen in section 7.2.1.

FIRST- AND SECOND-GENERATION METHODS 115

The Masamune auxiliaries (**71**) are prepared from (*R*)- or (*S*)-mandelic acid. The origin of the diastereofacial selectivity is much the same as it is in the Evans method (shown here for *R*).

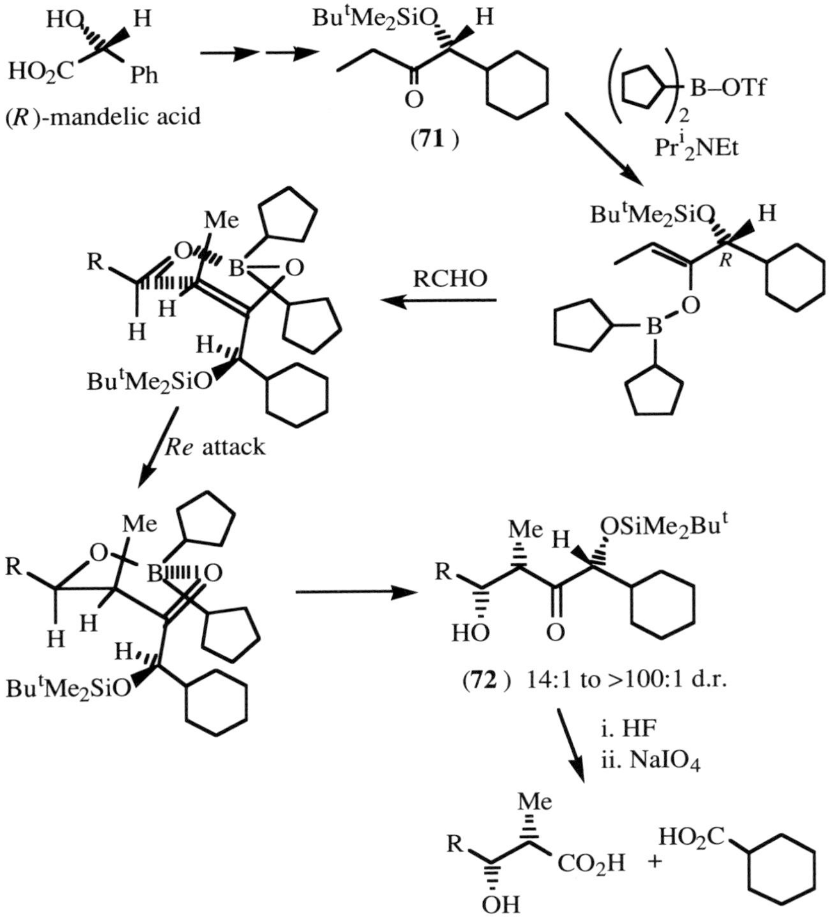

It can be seen that the diastereomeric transition state resulting from *Si* attack would be heavily hindered by interactions between the axial cyclopentyl group on the boron and the two bulky groups on the auxiliary.

The sequence is completed by removal of the silyl group in (**72**) with HF and cleavage of the α-hydroxyketone with periodate. A slight disadvantage of the method compared with that of Evans is that the chiral auxiliary is destroyed.

The question now arises: what if the aldehyde also bears a stereogenic centre? This is known as *double stereo-differentiation*,[25] or, for enantiomerically pure substrates, *double asymmetric induction*. We can distinguish two possibilities. In the first, the auxiliary on the enolate and the stereogenic centre on the aldehyde both favour the same diastereomer. Masamune calls this a 'matched pair'. In the second, the auxiliary and the stereogenic centre are opposed—a 'mismatched pair'. As one might expect, in the first case the d.e. is improved compared with single stereodifferentiation, whereas in the second the d.e. is reduced. An understanding of double stereo-differentiation is of great importance in the synthesis of polyether antibiotics in which diastereomeric purity is paramount.

X = SPh   A : B   3 : 2

X = (S, cyclohexyl, H, OSiMe₃)   A : B >100 : 1   (matched pair)

X = (R, cyclohexyl, Me₃SiO, H)   A : B   1 : 30   (mismatched pair)

As can be seen from this example, the achiral enolate gives a weak anti-Cram preference for the 3,4-*anti* product A. With the *S*-enolate (the matched pair) A becomes the exclusive product since the asymmetric inductions from the aldehyde side and from the chiral auxiliary are acting in the same sense. The enolate of the enantiomer gives predominantly the 3,4-*syn* product **B**, but with a lower diastereomeric ratio, since this is the mismatched pair.

All the examples discussed so far have been of *ethyl* enolates because simple methyl enolates give very poor diastereofacial selection. This problem can be avoided by the use of thiomethyl-substituted enolates. Raney nickel removes the sulphur to give the desired unsubstituted aldol (**73**).

As mentioned earlier, the preparation of *E*-enolates can present problems which render the *anti* aldols (**74**) inaccessible. Such products may be made efficiently with the trimethylsilyl derivative of *R*-triphenylethanediol as the chiral auxiliary.[26]

5.3.4 Asymmetric α-amino anions

The tetrahydroisoquinoline nucleus is found in literally thousands of alkaloids, many of which are of great medicinal importance. The tetrahydro-β-carboline nucleus is also widely distributed in the plant kingdom. Meyers has developed a valuable asymmetric synthesis of such systems which is based on an (S)-valine-derived chiral auxiliary, abbreviated to VBE.[27]

FIRST- AND SECOND-GENERATION METHODS 119

The methylene group adjacent to the nitrogen can be metalled with strong base to give an α-amino anion in which the lithium cation is chelated by the two heteroatoms of the chiral auxiliary. It is interesting to compare the alternative route to chiral 2-alkylated tetrahydroisoquinolines described in section 6.4.1.

The method works equally well with tetrahydro-β-carbolines, as shown by the asymmetric synthesis of the alkaloid yohimbone.

5.3.5 Chiral sulphoxides

The sulphoxide moiety is pyramidal and therefore a potential stereogenic centre capable of inducing asymmetry at adjacent centres. Furthermore, it is chemically versatile and can be removed under mild conditions. These features render the sulphoxide group attractive as a chiral auxiliary. Solladié has employed sulphoxide-stabilised anions in an asymmetric synthesis of 3-hydroxyesters (**76**).[28]

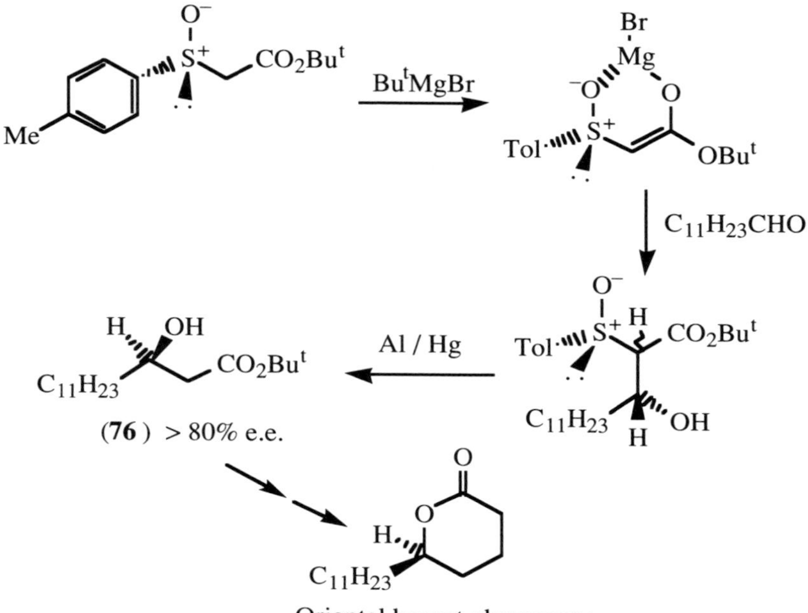

Oriental hornet pheromone

Two stereogenic centres are formed in the reaction between the anion (presumably a cyclic chelate) and the aldehyde but one is destroyed with the removal of the auxiliary, so its configuration is of no consequence. The nature of the aldehyde is important— acetylenic aldehydes give the *opposite* diastereomer (**77**).[29]

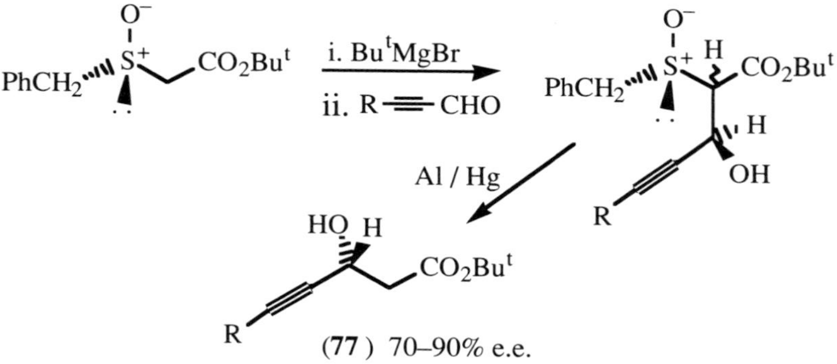

So far all the chiral compounds have been of the 'conventional' type, i.e. their stereogenic units have been tetrahedral carbon centres to which are attached four different substituents. The compounds (**78**) whose asymmetric synthesis is discussed below do not have a stereogenic centre: they possess axial chirality.

(*S*)-(**78**)   (*R*)-(**78**)

This synthesis, again due to Solladié,[30] uses a chiral sulphoxide group in a most ingenious way: it serves first to direct diastereoselective introduction of the ester group and then, by means of a stereospecific *syn* elimination, the double bond. The Grignard reagent (**79**) of the *trans*-4-substituted 1-bromomethyl-cyclohexane reacts with the enantiomerically pure (−)-(*S*)-menthyl *p*-toluene-sulphinate in an S_N2 displacement of the menthyloxy group, resulting in inversion at the sulphur stereogenic centre.

(*R*)-(**80**)   + Menthyl–OMgBr

The resulting (*R*)-sulphoxide (**80**) is then deprotonated by strong base at the methylene adjacent to the sulphur, and the resulting anion is carboxylated. Two diastereomers are formed (the S, R and the R, R) in unequal amounts, and these are methylated *in situ*. Simple column chromatography serves to separate the two diastereomers, which are then thermolysed.

The sulphoxide group undergoes clean *syn*-elimination with the result that the (S, R) diastereomer is converted exclusively to the (R)-enantiomer (**81**), while the (R, R) diastereomer gives its antipode (**82**). An alternative, third-generation approach to this type of compound is described in section 6.1.7.

Sulphoxides can also be used in the electrophilic mode (see section 5.4.1).

5.4 Electrophiles bearing chiral auxiliaries

5.4.1 Asymmetric Michael additions

The Meyers oxazoline auxiliary provides an efficient means of performing Michael addition to prochiral alkenes with excellent diastereofacial selectivity.[14] As shown below this allows convenient access to chiral β-alkylated acids (**83**). Note, however, that the facial selectivity is *opposite* to that observed in the alkylation of oxazoline anions (section 5.3.1).

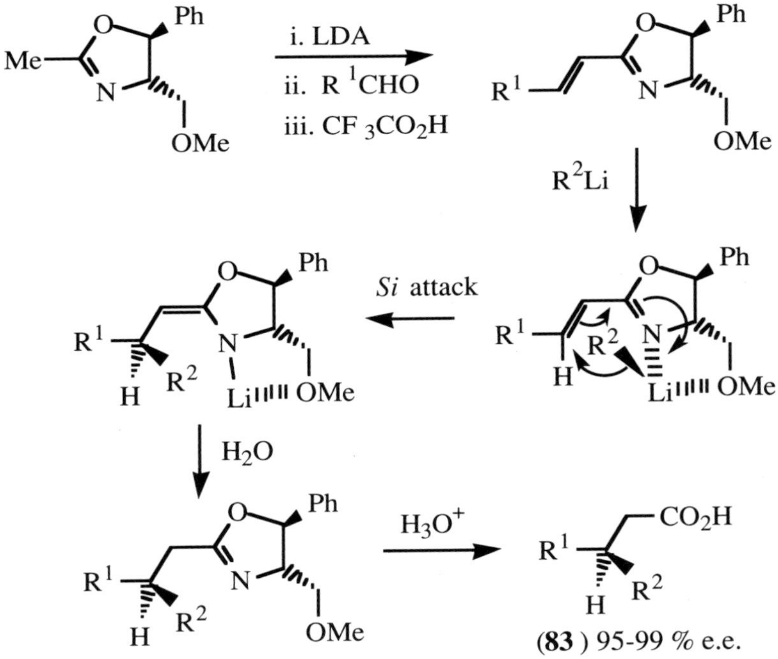

It has been known for some time that aromatic rings substituted with oxazolines are attacked in a Michael fashion by alkyllithium reagents. An asymmetric version of the reaction has been reported more recently by Meyers, and like so many other second-generation asymmetric syntheses, it relies on internal metal chelation to induce the necessary diastereofacial selectivity. Two typical examples are the conversion of (**84**) to (**85**) [31] and (**86**) to (**87**).[32] A further application of this chemistry is described in section 7.2.3.

The Meyers method is one of the very few asymmetric syntheses which allow the rational generation of axial chirality, as for example, in the chiral binaphthyl (**88**).[33] We will see an application of this axial asymmetric induction in section 7.2.4.

The sulphoxide group lends itself well to the asymmetric Michael addition,[34] since it both activates the double bond as an electrophile and directs the nucleophile to the face opposite the bulky aryl group as shown in the reactions of (**89**) and (**90**). The sulphoxide oxygen atom plays its customary role as a ligand in the rigid cyclic chelate.

Butenolides substituted with a chiral sulphoxide moiety are also good substrates, as shown by the short asymmetric synthesis of (−)-podorhizon (**91**), a lignan natural product.[35]

A very different approach to enantioselective Michael addition was developed by Seebach.[36] This involves conversion of the readily available (R)-3-hydroxybutyric acid (**92**) (see section 4.1.2) into a cyclic acetal (**93**) with pivalaldehyde. Unsaturation is then introduced resulting in disappearance of the original stereogenic centre. Fortunately, however, the information is recorded with total fidelity in the acetal centre. Nucleophilic attack on the double bond of (**94**) then occurs exclusively on the *ul (Si,R)* face. Within experimental limits, the (**93**) regenerated by hydrogenation has the same enantiomeric purity as it had at the beginning, and by cuprate addition a variety of products (**95**) containing the useful quaternary centre can be obtained. This approach has been called 'self regeneration of stereogenic centres' by Seebach.

A second stereogenic centre may be introduced by alkylation of the enolate, which takes place *anti* to the methyl group to give (**96**).[37] It is difficult to understand the enormous facial selectivity of these reactions, since the *t*-butyl group seems to be too far away to influence the reactivity. Calculations suggest that the effect is stereoelectronic in nature, and that the electrophilic carbon is slightly pyramidal.

The asymmetric Michael additions discussed so far have had a feature in common, namely that the double bond is incorporated in a ring. The greater challenge of designing a chiral auxiliary for acyclic systems has been met and solved by Oppolzer.[38] The auxiliary, derived from (+)-camphor, imposes severe restraints on the rotation of the acrylamide portion. It is believed that, with chelation of the amide and sultam oxygens to the magnesium cation, only the *S-cis* conformation (**97**) becomes energetically accessible. Attack then occurs with high selectivity from the *Si* face to give (**98**).

FIRST- AND SECOND-GENERATION METHODS

The intermediate enolate may be alkylated with high diastereoselectivity and electrophilic attack also occurs from the *Si* face to give (**99**).

A second example in which chelation to magnesium also plays a key role is the conversion of cinnamic acid to (**101**) by means of the ephedrine derivative (**100**).[39] Here addition takes place from the less hindered front face.

5.4.2 Chiral acetals

Acetals are unique as chiral auxiliaries since the prochiral carbon atom is *tetrahedral*, not trigonal, as in all other examples.[40] In the presence of a Lewis acid the acetal function becomes a powerful electrophile, capable of reacting with electron-rich double bonds. The origin of the selectivity is believed to be the preferential complexation of the Lewis acid with the less-hindered oxygen as in (**102**).[41] Reaction takes place by S_N2 displacement with inversion at the electrophilic carbon to give (**103**). Of interest in this synthesis of the lactone portion of mevinolin (**105**)[42] is the use of β-elimination to remove the chiral auxiliary and internal chelation to reduce the ketone (**104**) stereoselectively, as already described in section 5.2.4.

A spectacular application of chiral acetals is in the biomimetic polyene cyclisation to give the steroid skeleton (**106**) made famous by Johnson.[43] Here no fewer than seven new stereogenic centres are created in one step.

5.4.3 Asymmetric additions to carbonyl compounds

As mentioned earlier, the sulphoxide group is widely used in asymmetric synthesis. The final illustration of its use shows how it can direct the reduction of an adjacent carbonyl function as in (**107**) by complexation with the reducing agent.[44] It is particularly useful that either enantiomer of the final product (**108**) can be obtained depending on the reducing agent used.

The general and valuable Eliel method for the asymmetric synthesis of α-hydroxyaldehydes and other compounds is based on a 1,3-

oxathiane (**109**) derived from (*R*)-(+)-pulegone.[45] By analogy with dithianes, the oxathiane anion reacts with aldehydes as a chiral acyl anion equivalent. This gives exclusively the equatorial carbinol (**110**) but the diastereoselectivity at the alcohol centre is poor. However, when this is oxidised to (**111**) with the modified Swern reagent, Grignard reagents add to the carbonyl group from the *Re*-face with very high (> 90%) diastereomeric excess, as predicted by the Cram internally-chelated model. The chiral auxiliary is then removed with *N*-chloro-succinimide and silver nitrate, giving the sultine of the auxiliary (**112**) (which can be recycled) and the chiral α-hydroxyaldehyde (**113**). Since α-hydroxyaldehydes are rather unstable, these are generally transformed by standard chemistry into the diol, carbinol or α-hydroxyester.

In principle, either enantiomer is available simply by reversing the order of the introduction of groups R^1 and R^2.

5.5 Chiral auxiliaries in concerted reactions

5.5.1 Diels-Alder cycloaddition

There has been a vast amount of work directed towards making the Diels-Alder cycloaddition an enantioselective process, so only a small selection of the more modern methods will be shown here.[46]

The most popular location for the chiral auxiliary is on the dienophile, which is generally an acrylate ester of an optically active alcohol. The idea throughout is to favour addition of the diene to the *Re* or *Si* face of the acrylate by the steric blockade of the opposite face. One of the best acrylate auxiliaries is (−)-8-phenylmenthol, introduced by Corey.[47] The *Re* face is blocked here by the phenylisopropyl group leading to the formation of (**115**) as the major product.

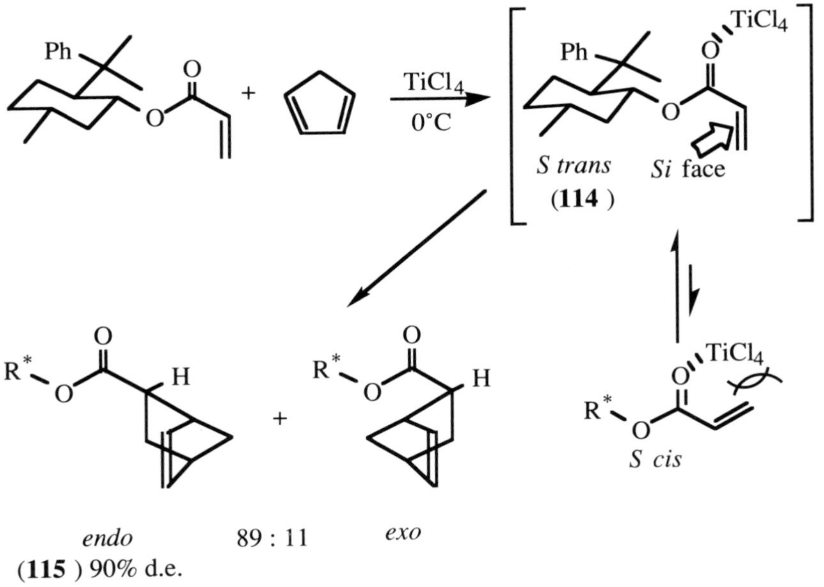

Almost invariably these cycloadditions are performed with Lewis acid catalysis. There are two main reasons for this: firstly, the reaction proceeds at low temperatures, increasing the facial selectivity and for cyclic dienes, the *endo* preference; secondly, the Lewis acid–

carbonyl complex is fixed in the *S-trans* conformation (**114**). Without the Lewis acid the acrylate has only a slight preference of 0.32 kcal mol^{-1} (1.34 kJ mol^{-1}) for the *S-trans* conformation, rendering facial selectivity meaningless. This explains why under non-catalytic conditions the d.e. is never better than 65%.

Many of the chiral auxiliaries described earlier serve excellently in asymmetric Diels-Alder reactions. The Evans oxazolidinones (**116**)[48] and the Oppolzer camphor sultams (**117**)[49] are exceptionally good in this regard and, as shown, actually give opposite products.

Naturally, the diastereoselective cycloaddition may be made intramolecular as illustrated by the reaction of (**118**).[50] In this case the real power of the method becomes apparent: we are forming four new stereogenic centres in a single step with completely defined relative and absolute configuration.

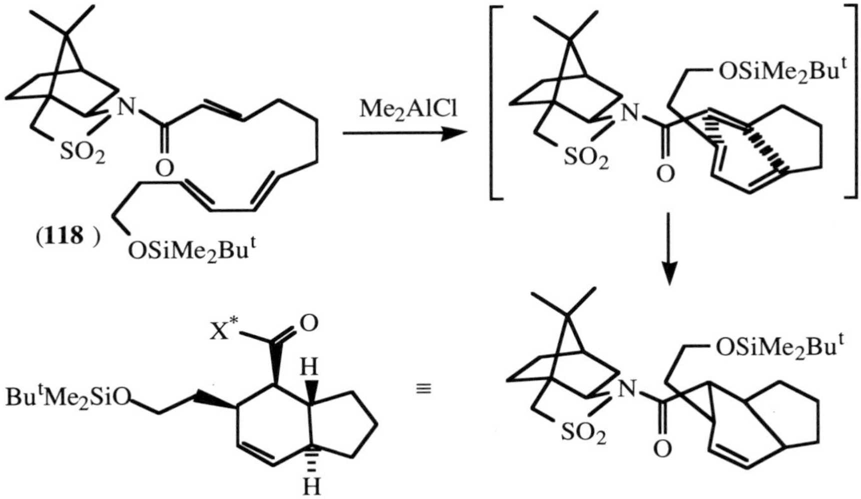

α-Hydroxyenones (**119**) are excellent chiral dienophiles.[51] The internal hydrogen bond places the stereogenic centre in a rigid five-membered ring, causing the bulky *t*-butyl group to block one face, in close analogy to the 'internal-chelation' Cram's rule. A catalyst is not necessary for reactive dienes in this case.

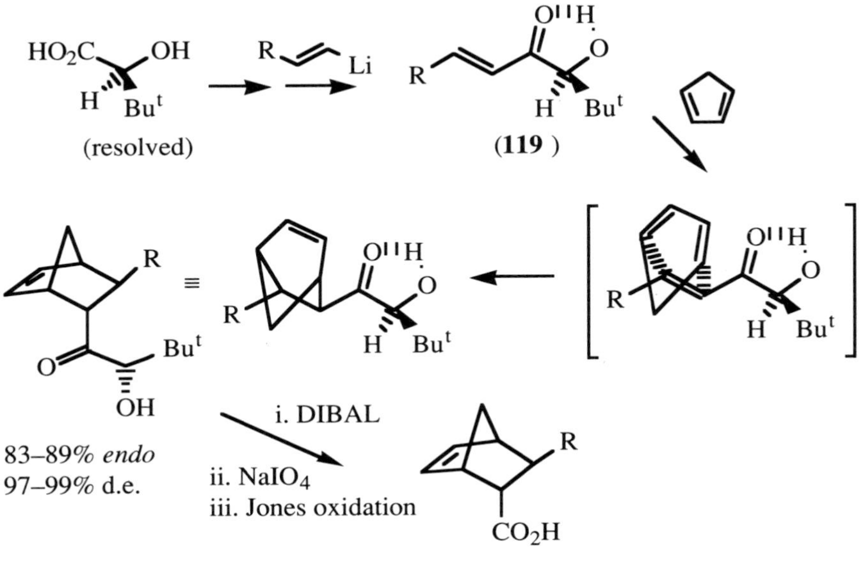

83–89% *endo*
97–99% d.e.

i. DIBAL
ii. NaIO$_4$
iii. Jones oxidation

Asymmetric Diels-Alder reactions with dienes bearing chiral auxiliaries have been studied less extensively. One example exploits some interesting cycloreversion chemistry.[52] The source of asymmetry for diene (**120**) is (*S*)-mandelic acid.

Another example of double asymmetric induction is seen in the reaction of (**120**) with the chiral α-hydroxyenone (**121**). The mismatched pair (*R*-diene, *S*-dienophile) gives the same product (**122**) (save with R* = *R*-O-methylmandelate), but with a somewhat lower d.e. of 94%. In other words the dienophile's facial selectivity overrides that of the diene.

5.5.2 The Claisen-Cope rearrangement

As a means of *transferring* the site of asymmetry the Claisen-Cope rearrangement is without comparison. Thus in the rearrangement of allyl vinyl ether (**123**) the original stereogenic centre is destroyed but two new centres are formed with retention of the overall chirality of the molecule to give (**124**).[53]

The rearrangement can be adapted to give a second-generation asymmetric synthesis by rendering the faces of one or other double bond diastereotopic as in (**125**) with a chiral auxiliary derived from (*S*)-valine.[54] 4-Pentenoic acids (**126**) substituted at C-2 are available in high e.e. by this method.

(**126**) > 93% e.e.

Like so many 6-electron concerted reactions, the Claisen-Cope rearrangement prefers to pass through a chair-like transition state (**127**), which explains its great predictability. The extension to 3-substituted pentenoates (**128**) is straightforward.[55]

For an account of an asymmetric ene reaction somewhat analogous to this example, see section 7.2.5.

References

1. H. Greuter, J. Dingwall, P. Martin and D. Bellus, *Helv. Chim. Acta,* 1981, **64**, 2812.
2. R. Zibuck, N.J. Liverton and A.B. Smith III, *J. Am. Chem. Soc.,* 1986, **108**, 2451.
3. M.R. Barbachyn and C.R. Johnson, in *Asymmetric Synthesis,* Vol. 4, J.D. Morrison and J.W. Scott, ed, Academic Press, Orlando, 1984, Chapter 2; C.R. Johnson, *Aldrichimica Acta,* 1985, **18**, 3.
4. C.R. Johnson and M. Barbachyn, *J. Am. Chem. Soc.,* 1982, **104**, 4290.
5. S. Hanessian, *Total Synthesis of Natural Products: The Chiron Approach,* Pergamon Press, Oxford, 1983.
6. D. Arnold, J.C.G. Drover and J.C. Vederas, *J. Am. Chem. Soc.,* 1987, **109**, 4649.
7. D.M. Floyd, A.W. Fritz, J. Pluscec, E.R. Weaver and C.M. Cimarusti, *J. Org. Chem.,* 1982, **47**, 5160.
8. R.C. Anderson and B. Fraser-Reid, *J. Org. Chem.,* 1985, **50**, 4781.
9. S. Hanessian, *Aldrichimica Acta,* 1989, **22**, 3.

10. D.R. Williams, B.A. Barner, K. Nishitani and J.G. Phillips, *J. Am. Chem. Soc.*, 1982, **104**, 4708.
11. G. Wess, K. Kesseler, E. Baader, W. Bartmann, G. Beck, A. Bergmann, H. Jendralla, K. Bock, G. Holzstein, H. Kleine and M. Schierer, *Tetrahedron Lett.*, 1990, **31**, 2545.
12. G. Cainelli, M. Pannuzio, G. Giacomini, G. Martelli and G. Spunta, *J. Am. Chem. Soc.*, 1988, **110**, 6879; P. Andreoli, G. Cainelli, M. Pannuzio, E. Baudrin, G. Martelli and G. Spunta, *J. Org. Chem.*, 1991, **56**, 5984.
13. D. Enders, in *Asymmetric Synthesis*, Vol. 3, J.D. Morrison ed., Academic Press, Orlando, 1984, Chapter 4.
14. K.A. Lutomski and A.I. Meyers, in *Asymmetric Synthesis*, Vol. 3, J.D. Morrison ed., Academic Press, Orlando, 1984, Chapter 3.
15. A.I. Meyers, B.A. Lefker and R.A. Aitken, *J. Org. Chem.*, 1986, **51**, 1936. For a review of the many other useful applications of these chiral bicyclic lactams in asymmetric synthesis see: D. Romo and A.I. Meyers, *Tetrahedron*, 1991, **47**, 9503.
16. U. Schöllkopf, T. Tiller and J. Bardenhagen, *Tetrahedron*, 1988, **44**, 5293 and earlier references cited therein.
17. H.J. Zeiss, *Tetrahedron Lett.*, 1987, **28**, 1255.
18. R.M. Williams, *Synthesis of Optically Active α-Amino Acids*, Pergamon Press, Oxford, 1989. See also M.J. O'Donnell (ed.), *Tetrahedron Symposium in Print, Tetrahedron*, 1988, **44**, 5253ff.
19. D.A. Evans, M.D. Ennis and D.J. Mathre, *J. Am. Chem. Soc.*, 1982, **104**, 1737; D.A. Evans, M.M. Morrissey and R.L. Dorow, *J. Am. Chem. Soc.*, 1985, **107**, 4346; D.A. Evans, T.C. Britton, R.L. Dorow and J.F. Dellaria, *J. Am. Chem. Soc.*, 1986, **108**, 6395.
20. S.G. Davies, *Pure Appl. Chem.*, 1988, **60**, 131; *Aldrichimica Acta*, 1990, **23**, 31.
21. C.H. Heathcock, in *Asymmetric Synthesis*, Vol. 3, J.D. Morrison ed., Academic Press, Orlando (1984), Chapter 2.
22. D.B. Collins, J.H. McDonald and W.C. Still, *J. Am. Chem. Soc.*, 1980, **102**, 2120.
23. D.A. Evans, J.V. Nelson and T.R. Taber, *Top. Stereochem.*, 1982, **13**, 2; D.A. Evans, *Aldrichimica Acta*, 1982, **15**, 23; D. A. Evans and L.R. McGee, *J. Am. Chem. Soc.*, 1981, **103**, 2876.
24. S. Masamune, in *Organic Synthesis Today and Tomorrow*, B.M. Trost, ed., Pergamon Press, 1981; S. Masamune, W. Choy, F.A.J. Kerdesky and B. Imperiali, *J. Am. Chem. Soc.*, 1981, **103**, 1566.
25. S. Masamune, W. Choy, J.S. Petersen and L.R. Sita, *Angew. Chem., Int. Ed. Engl.*, 1985, **24**, 1.
26. M. Brown and H. Sacha, *Angew. Chem., Int. Ed. Engl.*, 1991, **30**, 1318.
27. A.I. Meyers, *Tetrahedron*, 1992, **48**, 2589; *Aldrichimica Acta*, 1985, **18**, 59.
28. G. Solladié, in *Asymmetric Synthesis*, Vol. 2, J.D. Morrison ed., Academic Press, Orlando, 1983, Chapter 6.
29. G. Solladié, G. Frechon and G. Demailly, *Nouv. J. Chem.*, 1985, **9**, 21.
30. G. Solladié, R. Zimmerman, R. Bartsch and H.M. Walborsky, *Synthesis*, 1985, 662.
31. A.I. Meyers and B.A. Barner, *J. Org. Chem.*, 1986, **51**, 120.
32. A.I. Meyers and T.O. Oppenlaender, *J. Chem. Soc., Chem. Commun.*, 1986, 920.
33. A.I. Meyers and R.J. Himmelsbach, *J. Am. Chem. Soc.*, 1985, **107**, 682.
34. G.H. Posner, in *Asymmetric Synthesis*, Vol. 2, J.D. Morrison ed., Academic Press, Orlando, 1983, Chapter 8; G.H. Posner, L.L. Frye and M. Hulce, *Tetrahedron*, 1984, **40**, 1401.
35. G.H. Posner, T.P. Kogan, J.R. Haines and L.L. Frye, *Tetrahedron Lett.*, 1984, **25**, 3627.

36. D. Seebach, J. Zimmermann, U. Gysel, R. Ziegler and T.-K. Ha, *J. Am. Chem. Soc.*, 1988, **110**, 4763.
37. D. Seebach and J. Zimmermann, *Helv. Chim. Acta*, 1986, **69**, 1147.
38. W. Oppolzer, *Tetrahedron,* 1987, **43**, 1969; *Pure Appl. Chem.*, 1990, **62**, 1241.
39. T. Mukaiyama and N. Iwasawa, *Chem. Lett.*, 1981, 913.
40. A. Alexakis and P. Mangeney, *Tetrahedron: Asymmetry,* 1990, **1**, 477.
41. S.E. Denmark and N.G. Almstead, *J. Org. Chem.*, 1991, **56**, 6458.
42. W.S. Johnson, A.B. Kelson and J.D. Elliot, *Tetrahedron Lett.*, 1988, **29**, 3757.
43. D. Gray, W.S. Johnson and U. Schubert, *J. Org. Chem.*, 1989, **54**, 4731.
44. G. Solladié, G. Greck, G. Demailly and A. Solladié-Cavallo, *Tetrahedron Lett.*, 1982, **23**, 5047.
45. E.L. Eliel, in *Asymmetric Synthesis,* Vol. 2, J.D. Morrison ed., Academic Press, Orlando, 1983, Chapter 5; J. Lynch and E.L. Eliel, *J. Am. Chem. Soc.*, 1984, **106**, 2943.
46. For reviews see: L.A. Paquette, *Asymmetric Synthesis,* Vol. 3, J.D. Morrison ed., Academic Press, Orlando, 1984, Chapter 7; W. Oppolzer, *Angew. Chem., Int. Ed. Engl.*, 1984, **23**, 876.
47. E.J. Corey and H.E. Ensley, *J. Am. Chem. Soc.*, 1975, **97**, 6908.
48. D.A. Evans, K.T. Chapman and J. Bisalia, *J. Am. Chem. Soc.*, 1984, **106**, 4261.
49. W. Oppolzer, C. Chapuis and G. Bernardinelli, *Helv. Chim. Acta*, 1984, **67**, 1397.
50. W. Oppolzer, D. Dupuis, G. Poli, T.M. Raynham and G. Bernardinelli, *Tetrahedron Lett.*, 1988, **29**, 5885.
51. S. Masamune, L.A. Reed III, J.T. Davis and W. Choy, *J. Org. Chem.*, 1983, **48**, 4441.
52. B.M. Trost, D. O'Krongly and J. Belletire, *J. Am. Chem. Soc.*, 1980, **102**, 7595.
53. R.K. Hill, in *Asymmetric Synthesis*, Vol. 3, J.D. Morrison ed., Academic Press, Orlando, 1984, Chapter 8.
54. M.J. Kurth, O.H.W. Decker, H. Hope and M.D. Yannick, *J. Am. Chem. Soc.*, 1985, **107**, 443.
55. M.J. Kurth and O.H.W. Decker, *J. Org. Chem.*, 1985, **50,** 5769.

6 Third- and fourth-generation methods: asymmetric reagents and catalysts

S.N. KILENYI and R.A. AITKEN

We discuss in this chapter some third- and fourth-generation methods of asymmetric synthesis. As the reader will recall, these are distinguished from the first- and second-generation methods by the fact that the asymmetry derives from a reagent or catalyst, rather than from the starting material or a chiral auxiliary.

$$S \xrightarrow{R^* \text{ or } R / \text{cat.}^*} P^*$$

The advantages over first- and second-generation methods are two-fold: the choice of starting material is far wider, since it need no longer come from the chiral pool, and there is no need to dedicate two extra steps to the installation and removal of a chiral auxiliary.

6.1 C–C bond-forming reactions

6.1.1 Asymmetric alkylation

The first asymmetric reaction discussed here, of the fourth-generation type, is based on earlier work by Wynberg and was reported by a group at Merck who were interested in the industrial synthesis of the diuretic indacrinone.[2] The chiral catalyst is a quaternary salt of a

---

A note to the reader: due to the immense quantity of work in this field, and the limited space available, the treatment given here may seem rather superficial and empirical. Furthermore, in many cases the mechanisms involved are not yet fully understood. The original literature and *Asymmetric Synthesis*[1] should be consulted for greater depth and detail.

cinchona alkaloid,[3] which functions as a phase-transfer catalyst. The overall reaction is as follows:

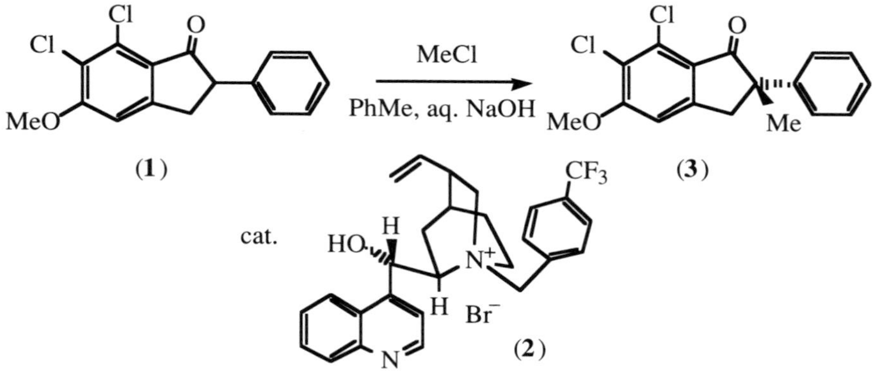

The phase-transfer catalyst (**2**), derived from natural cinchonine, reacts with NaOH in the aqueous layer to give the quaternary ammonium hydroxide which then moves across into the organic layer and deprotonates the (racemic) starting ketone (**1**) to form the enolate, which is then alkylated. The product (**3**), a precursor of indacrinone, is formed in 95% yield and 92% e.e. The high e.e. derives from the fact that the enolate and quaternary ammonium ion form a tight ion pair in which the two enantiotopic faces of the enolate are no longer equivalent. The electrophile is then obliged to attack from the outer, less hindered face. The conditions shown above are the result of extensive optimisation of all the reaction parameters including the concentrations, temperature, solvent, nature of the methyl electrophile, and nature of the substituent on the benzyl group of the catalyst. Only when all these are right is the highest e.e. achieved. The reaction is of limited scope but other alkyl halides can be used as shown in the complete synthesis of a drug molecule related to (**3**) which is described in section 7.4.1. The *cinchona* alkaloids are also good catalysts for asymmetric Michael addition of indanone anions as shown below.

6.1.2 Asymmetric Michael reaction

In a continuation of the work described above, the Merck group also found (**2**) to be an efficient catalyst for the asymmetric Michael addition of indanone (**4**) to methyl vinyl ketone (**5**).[4] The product (**6**) was formed in 95% yield and 80% e.e. Interestingly the diastereomer

of (**2**) derived from cinchonidine gave the enantiomer of (**6**) in good yield but only 40% e.e.

An earlier and closely related asymmetric Michael reaction was reported by Wynberg and Hermann.[5] The ketoester (**7**) in this case is acidic enough to be deprotonated by quinine (**8**), whose structure is similar to cinchonine but enantiomeric at the two key stereogenic centres. In the non-polar tetrachloromethane solvent the enolate and protonated quinine form a tight ion-pair very similar to before. The enolate then reacts largely from the *Re* face to give the *R* product (**9**).

Octahedral complexes of transition metal ions with bidentate ligands can display chirality:

$M = Fe^{3+}, Co^{2+}$, etc.

Δ   Λ

If one of the ligands contains one or more stereogenic centres, then the two enantiomers Δ and Λ become diastereomers, and therefore distinguishable chemically. By the correct choice of stereogenic ligand, one can favour formation of the Δ or Λ form to the exclusion of the other (see section 6.4 for another example of this). A cobalt(II) complex of this type has been used as a catalyst in an asymmetric Michael reaction.[6] The asymmetry derives from the chiral ligand (+)-S,S-1,2-diphenyl-1,2-diaminoethane. On reaction with Co(II) (acac)$_2$ the Δ-diastereomer is formed preferentially. Only 3% of this Δ complex of cobalt (**11**) suffices to give a 66% e.e. in the following Michael reaction. The reaction proceeds by the replacement of the acetylacetone ligands by the ketoester (**10**). It is this new complex (**13**) which adds to the Michael acceptor (only one of the two ketoester ligands is shown, and the chiral diamine is simplified). Once again we see the principle of enantiotopic differentiation at work: the enolate of the ketoester is flat, and both faces are equivalent until it complexes with the chiral catalyst.

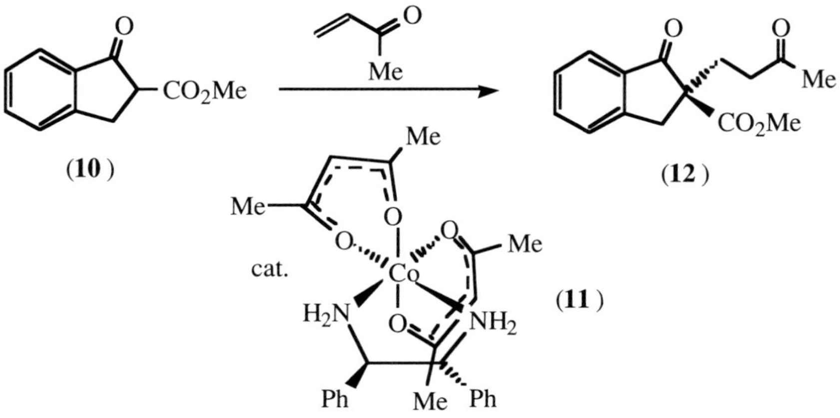

(10) + acrylate (Me) →(cat. 11)→ (12)

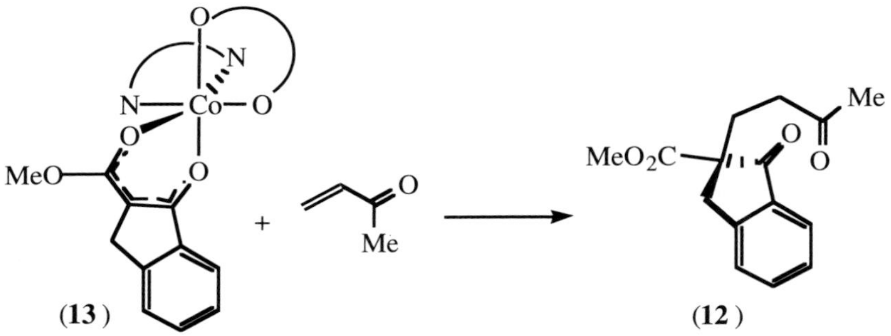

(11) Co catalyst with chiral diamine ligand

(13) + acrylate → (12)

The method described above results in asymmetric induction at the nucleophile. Asymmetry can also be induced at the Michael acceptor. An example of this involving C–S bond formation is shown below.[7] (For the sake of consistency, this reaction is considered with the C–C bond forming reactions.)

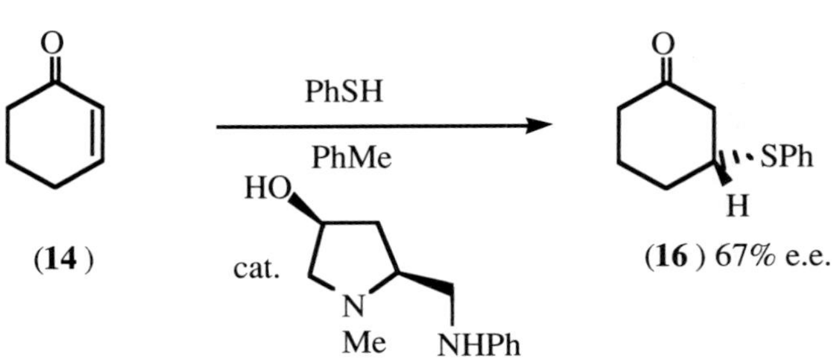

(14) + PhSH →(PhMe, cat. 15)→ (16) 67% e.e.

The chiral catalyst (**15**) is derived from natural *S,S*-hydroxyproline, and serves to deprotonate the thiophenol. A tight ion pair is formed with (**14**), rather reminiscent of that in asymmetric phase-transfer catalysis. In a non-polar solvent such as toluene, solvation is at a premium, so one would expect the hydroxy of the catalyst to hydrogen bond to the carbonyl of the enone (**17**).

Attack from the *Re*-face is favoured because attack from the *Si* face results in severe steric hindrance between the enone ring and the aryl ring of the catalyst. This asymmetric reaction has limitations, as one would expect: aliphatic thiols do not react well, being insufficiently acidic, and cyclopentenones give poor enantiomeric excesses.

6.1.3 Asymmetric nucleophilic additions to carbonyl compounds

The addition of an organometallic reagent to a carbonyl group results in the formation of a new stereogenic centre. A considerable amount of work has been devoted to making this process asymmetric.[3] The best results so far have been obtained with dialkylzinc reagents complexed with chiral ligands.[8] Thus, in the method of Noyori, a dialkylzinc (usually diethylzinc) is added to an aldehyde in the presence of a catalytic amount (1 mol %) of a chiral aminoalcohol (**18**) derived from (–)-camphor. The dialkylzinc reacts with the aminoalcohol to form a chelate (**19**) with loss of one alkyl group as the alkane. Zinc alkyls are coordinatively unsaturated and exist as bridged dimers. It is believed to be the 'dimer' (**20**) formed between the aminoalcohol/zinc alkyl complex and another molecule of dialkylzinc which actually reacts with the aldehyde to give the secondary alcohol (**21**). The enantiomeric excesses are excellent ($\geq 98\%$), and the reaction can be applied to a number of aromatic and aliphatic aldehydes.

A closely related method uses a polymer-bound (−)-ephedrine catalyst (**22**). The enantiomeric excesses are, however, rather less impressive (33–89%).[9]

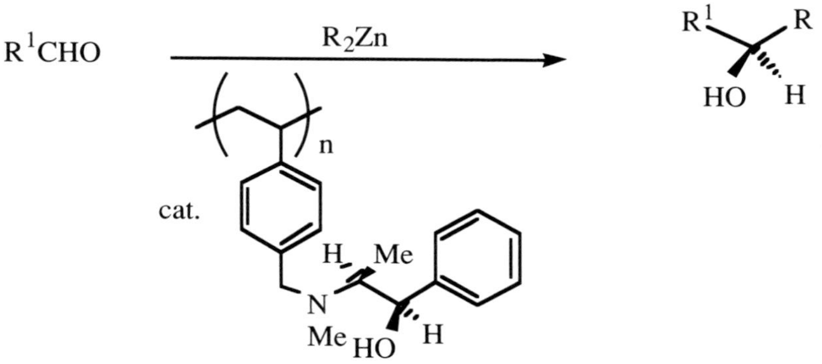

6.1.4 Asymmetric [2 + 2] cycloadditions[10]

Ketene undergoes rapid stepwise [2+2] cycloaddition to the carbonyl group of chloral (trichloroacetaldehyde) in the presence of a catalytic amount of an amine to afford the β-propiolactone (**23**):

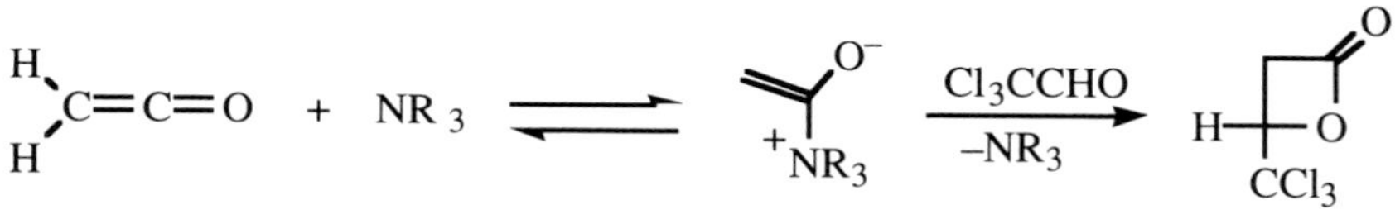

If this reaction is conducted under optimum conditions (toluene, −50°) in the presence of 1–2 mol % quinidine as chiral basic catalyst, the β-propiolactone is obtained with 98% e.e. as the (*R*)-enantiomer (**25**). Use of quinine leads to the (*S*)-enantiomer (**24**) in 76% e.e. To

THIRD- AND FOURTH-GENERATION METHODS

confirm the assignments of absolute stereochemistry, the propiolactones were converted into the known (S)- and (R)-malic acids

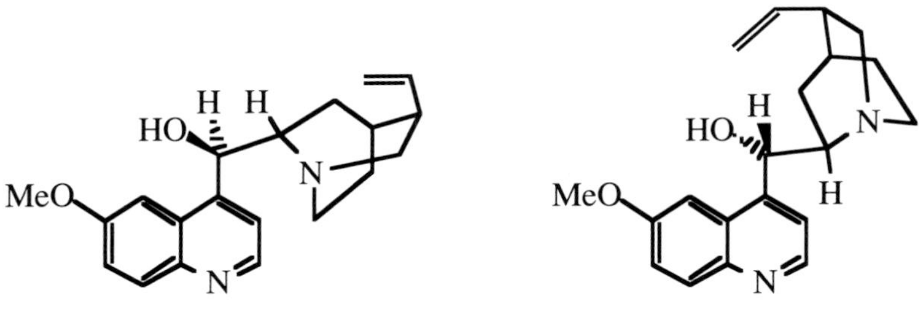

quinine

quinidine

However, this is a good example of the dangers which face the unwary in asymmetric synthesis, since it was later discovered that the hydrolysis step is actually accompanied by efficient *inversion of absolute configuration* at the stereogenic centre. Thus it is the (S)-product from quinine which gives the (R)-malic acid and vice versa as shown. Regardless of this complication, the method allows convenient access to either enantiomer of the synthetically useful malic acid on a commercial scale.

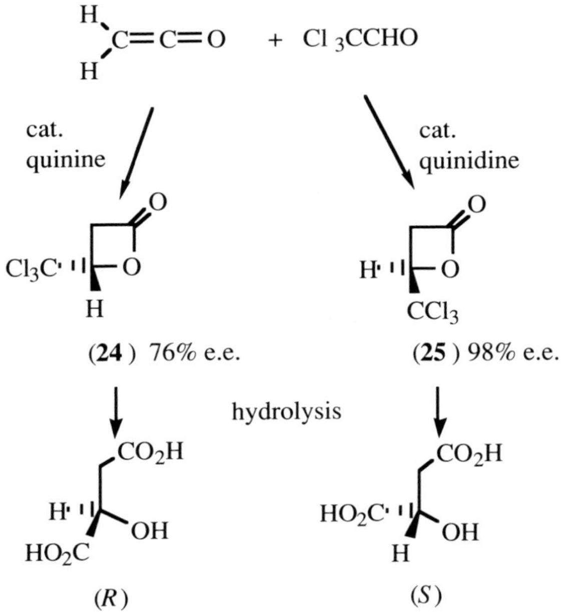

$H_2C=C=O$ + Cl_3CCHO

cat. quinine

cat. quinidine

(24) 76% e.e.

(25) 98% e.e.

hydrolysis

(R)

(S)

152   ASYMMETRIC SYNTHESIS

Electron-rich and electron-poor alkenes can react with each other in a [2 + 2] fashion to give cyclobutanes with Lewis acid catalysis. In the example below, the chiral catalyst (**27**) is a complex of titanium(IV) with a ligand derived from (−)-tartaric acid. In the presence of 10 mol% of this catalyst, a stepwise [2 + 2]-cycloaddition occurs to (**26**) affording the cyclobutane (**28**) with an e.e. of > 98%. The product was converted in a few steps to a carbocyclic analogue of the nucleoside antibiotics oxetanocins A and G.[11]

6.1.5 Asymmetric Diels-Alder reaction[10]

The Diels-Alder reaction has not escaped the attention of chemists interested in asymmetric synthesis. A binaphthyl-based chiral aluminium complex (**30**) catalyses the hetero Diels-Alder cycloaddition between benzaldehyde and the Danishefsky diene (**29**), providing the dihydropyrone (**32**) with excellent selectivity after acid hydrolysis of the initial adduct (**31**).[12] (We will see further applications of chiral binaphthyl ligands in section 6.4.)

Another example, this time of a carbocyclic Diels-Alder reaction, is shown below.[10] The catalyst (**33**) is, of course, derived from (−)-menthol. In both cases the diene reacts with the complex of the chiral Lewis acid and the dienophile, whose two enantiotopic faces are no longer equivalent.

6.1.6 Crotylboranes

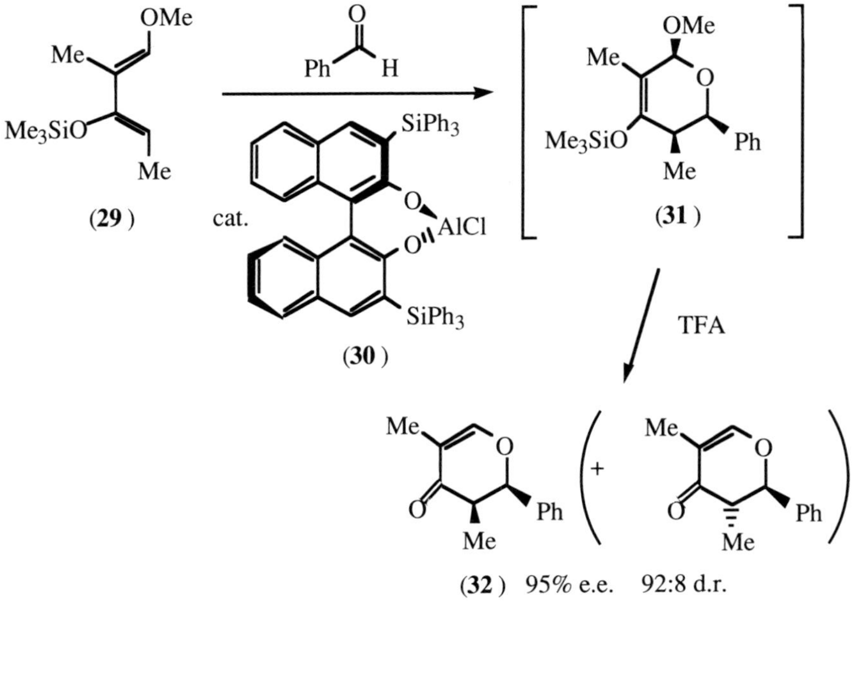

A valuable asymmetric route to homoallylic alcohols exploits the remarkably versatile chemistry of boranes.[13] The (Z)- and (E)-crotylboranes (**37**) and (**38**) are prepared by reaction of the corresponding (Z)- and (E)- crotyl potassiums with (+)- or (−)-methoxydiisopino-campheylborane (**36**), (also known as Ipc_2BOMe), which in its turn is derived by hydroboration of (+)- or (−)-α-pinene (**35**).

Aldehydes react smoothly with these crotylboranes *via* a six-membered Zimmerman–Traxler type transition state (**39**), giving homoalkylic alcohols (**40**) with high enantioselectivity and essentially complete diastereoselectivity.

By the appropriate choice of alkene geometry and chirality of pinene, it is possible to make either enantiomer of either diastereomer at will. For further applications of chiral boranes, see section 6.4.

6.1.7 Asymmetric formation of alkene double bonds

So far, all the chiral compounds prepared have been of the 'conventional' type, i.e. with tetrahedral stereogenic centres. The compounds whose asymmetric synthesis is discussed below do not have a stereogenic centre: the 4-substituted alkylidenecyclohexanes possess axial chirality.

A second generation route to this class of compound using sulphoxide chiral auxiliaries has already been described in section 5.3.5. The third-generation method here is an asymmetric variant of the Wittig reaction.[14] The asymmetric reagent is derived from (S, S)- or (R, R)-1,2-diaminocyclohexane (note the C_2 symmetry). Deprotonation and reaction with a 4-substituted cyclohexanone leads to one or other enantiomer of the axially chiral alkylidenecyclohexane (**43**) with a good e.e. of 70–90%. The origin of this selectivity is not yet fully understood.

6.2 Chiral acids and bases

The asymmetric reactions discussed below have one feature in common, namely that the substrates have two reaction sites and a plane of symmetry, and the chiral reagent reacts preferentially with one of the two enantiotopic groups.

The first example involves the asymmetric protonation[15] of a hydroxydicarboxylate. When the disodium salt of 4-hydroxypimelic acid (**44**) is protonated with (*S*)-camphorsulphonic acid (**45**) under optimal conditions (3.5 mM in ethanol, –78°) the (*S*)-lactone (**46**) is formed with an enantiomeric excess of 94%. The authors believe that the reaction proceeds by enantiotopic protonation of the *Re*-carboxylate, followed by lactonisation.

A deracemisation is defined as the conversion of a racemic compound into an enantiomerically-enriched form. (Note: this is **not** a resolution, in which the two enantiomers are separated: in a deracemisation, one enantiomer is converted into the other.) An example of this is provided by the following asymmetric reaction.[16]

Racemic benzoin is converted to its dianion (**47**), which of course is achiral, being planar. On enantioselective protonation with

(2R, 3R)-O,O-dipivaloyltartaric acid (**48**), the (S)-enantiomer (**49**) is formed with an e.e. of up to 80%.

Deprotonation of prochiral ketones with chiral amide bases such as (**50**), derived from the readily available enantiomerically pure 1-phenylethylamine, provides an efficient asymmetric synthesis of silyl enol ethers (**51**).

This enantioselective deprotonation is not limited to ketones flanked by methylene groups as illustrated by the reaction of *cis*-2,6-dimethylcyclohexanone with chiral base (**52**) to give (**53**). These asymmetric deprotonations have great synthetic potential.[17]

Perhaps the most celebrated example of enantiotopic differentiation is that due to Hajos and Parrish.[18] The substrate is a triketone (**54**) with a plane of symmetry, and the source of asymmetry is the natural amino acid (S)-proline which, being amphoteric, can be classed as both an acid and a base. Simply stirring a DMF solution of (**54**) and 3 mol % (S)-proline at room temperature results in quantitative conversion to the ketol (**55**), with an e.e. of 93.4%. One recrystallisation raises this to almost 100%.

Dehydration of the ketol affords the enedione (**56**), an important intermediate in steroid chemistry. The corresponding reaction of (**57**) can be used to make the famous Wieland–Miescher ketone (**58**) in optically active form.[19]

The origin of this high enantiotopic differentiation is still not entirely clear. Both the secondary amine and the carboxylic acid are necessary, since *N*-methylproline and prolinol give little or no asymmetric induction.

6.3 Asymmetric oxidation methods

6.3.1 Asymmetric epoxidation of alkenes

Regarded by many as one of the most important advances of the 1980s, the Katsuki–Sharpless asymmetric epoxidation[20] is a powerful fourth-generation method of synthesising epoxylalcohols, which are versatile intermediates, with predictable absolute configuration and very high enantiomeric excess.

The reaction is normally performed at low temperatures (–30 to 0°) in methylene chloride, and is *catalytic* in the chiral component diethyl or diisopropyl tartrate (DET or DIPT), and in titanium tetra-isopropoxide, provided water is rigorously excluded: 4 Å molecular sieves may be added to ensure this. Both enantiomers of tartaric acid are commercially available, allowing the synthesis of either enantiomer of the epoxylalcohol. The key to the remarkable enzyme-like enantioselectivity lies in the complex formed from the

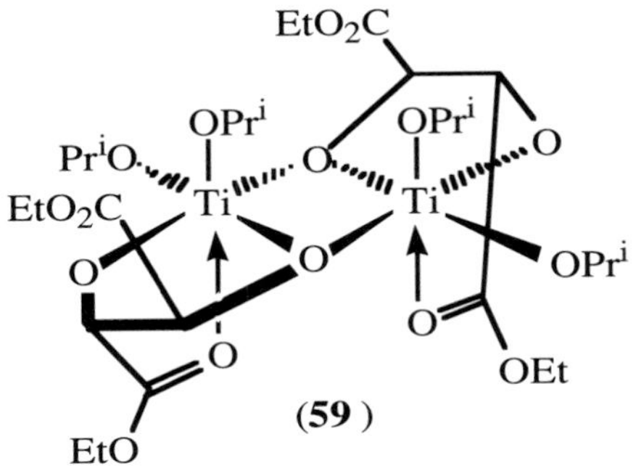

THIRD- AND FOURTH-GENERATION METHODS

(59)

titanium salt and the tartrate. It is believed to have the structure (**59**). Under the reaction conditions ligand exchange occurs rapidly with the actual oxidant (Bu^{t}OOH) and the allylic alcohol. In the highly asymmetric environment of the binuclear titanium complex, attack of the complexed hydroperoxide is forced to occur from the *Si*- face (with (+)-DET) or *Re*- (with (–)-DET), due to the presence of the bulky ester groups. It should be noted, however, that the exact nature of the substrate/catalyst complex is still controversial.

A typical example of the conditions used is provided by the reaction of undec-2-en-1-ol to afford epoxide (**60**) in 96% yield with an e.e. of 95%.

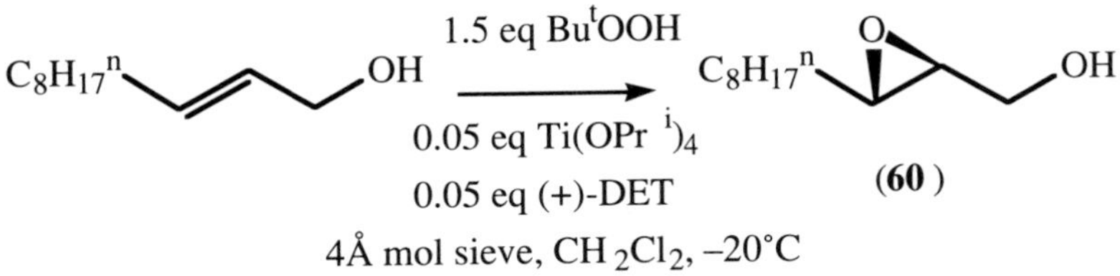

$C_8H_{17}^n$ ⁀⁀⁀OH

1.5 eq Bu^{t}OOH
0.05 eq Ti(OPr^i)$_4$
0.05 eq (+)-DET
4Å mol sieve, CH$_2$Cl$_2$, –20°C

$C_8H_{17}^n$ ⁀⁀⁀OH (**60**)

There are often problems encountered in the isolation of low molecular weight epoxy-alcohols such as the parent compound glycidol, due to their water solubility. Such difficulties can be avoided by derivatisation *in situ* as the *p*-toluenesulphonate or *p*-nitrobenzenesulphonate ester and this procedure is illustrated in the application of the Sharpless oxidation to prepare (–)-propranolol described in section 7.4.2.

An interesting variant on this scheme involves kinetic resolution (see section 4.2.4): a racemic secondary allyl alcohol (**61**) is epoxidised with the Sharpless reagent. Only **one** enantiomer reacts. It is then a simple matter to separate the enantiomerically enriched starting material (**62**) from the epoxylalcohol (**63**).

6.3.2 Asymmetric oxidation of sulphides

Although it is well known that the sulphoxide group can be a stereogenic centre, general methods of synthesising enantiomerically enriched sulphoxides are very rare, despite the synthetic potential of this function. A new and important asymmetric synthesis of sulphoxides directly from the corresponding sulphides is based on a variant of the Sharpless oxidation.[21]

As can be seen from the above example, there are two important differences from the normal Sharpless procedure: first, because the sulphoxide is a good ligand for titanium, the reaction must be performed with a *stoichiometric* quantity of tartrate/titanium tetraisopropoxide; second, one equivalent of water is necessary for good e.e.s. If water is strictly excluded, the enantiomeric excess drops

THIRD- AND FOURTH-GENERATION METHODS   163

to nearly zero, the exact opposite of the situation which applies in the Sharpless epoxidation. Clearly, the nature of the titanium complex, believed to be mononuclear, is different. The reaction is tolerant of a wide variety of functional groups (e.g. ester, nitro, pyridyl) and can be used to make dialkyl sulphoxides asymmetrically, although as shown by the example (**65**), the e.e.s are less impressive than those with alkyl tolyl sulphides (**64**).

$$Bu^t-S-Me \xrightarrow[CH_2Cl_2/H_2O]{Bu^tOOH \quad (+)\text{-DET}/Ti(OPr^i)_4} Bu^t-S(=O)\text{-}Me$$

(**65**) 53% e.e.

6.3.3 Asymmetric dihydroxylation

The familiar *cis*-dihydroxylation of alkenes with OsO_4 or $KMnO_4$ creates two new stereogenic centres. Normally, this is of no consequence since in most applications the diol is cleaved or the alkene has a plane of symmetry.

$$\underset{R^1\;\;R^2}{R^3\;\;R^4}\text{C=C} \xrightarrow{OsO_4 \text{ or } KMnO_4} R^3(HO)(R^1)C-C(OH)(R^2)R^4$$

An asymmetric version of this reaction would be of great interest, since it would complement the asymmetric epoxidation. This has in fact been achieved by Sharpless. The source of asymmetry this time is a chiral tertiary amine, which forms a complex with osmium tetroxide. Taking their cue from the work of Wynberg (see section 6.1.1) Sharpless and co-workers discovered once again that chiral amines (**66**) and (**67**) derived from the *cinchona* alkaloids quinine and quinidine respectively performed well, affording enantiomeric excesses of 50–90%.

The reaction was originally performed using a stoichiometric amount of the oxidant OsO_4, with attendant problems of toxicity and cost.[22] The appeal of the method was greatly enhanced by the discovery that the asymmetric oxidation worked just as well under the more modern catalytic conditions, with only a trace of osmium tetroxide and N-methylmorpholine N-oxide (**68**) as the reoxidant.[23] The method is tolerant of a wide variety of functional groups (esters, allylic halides) and, unlike the Sharpless oxidation, has no need of an allylic hydroxyl group as an internal ligand.[24]

A limitation of the catalytic procedure, unlike the stoichiometric, is that hindered and 1,1-disubstituted alkenes give poor e.e.s. However, simply adding the alkene slowly to the reaction mixture results in greatly improved optical yields, even with these substrates.[25] Furthermore, the reaction is then actually *faster*.

THIRD- AND FOURTH-GENERATION METHODS

	e.e.
Stoichiometric (R = CH$_3$CO, 1.1 eq OsO$_4$, toluene)	79%
Catalytic (R = p-Cl-C$_6$H$_4$CO, cat. OsO$_4$, (**68**), aq acetone)	8%
Catalytic (as above but with slow addition of alkene)	78%

This strange observation is explained by the fact that there are two catalytic cycles operating, the first, predominant at low alkene concentrations, being fast and highly enantioselective, and the second, slow and almost unselective. By minimising the stationary concentration of alkene, the second catalytic cycle is suppressed.

6.3.4 Chiral oxaziridines and their uses[26]

The final method discussed in this section on asymmetric oxidation relies on the properties of a rather rare functional group, the oxaziridine, which is a potential tautomer of the nitrone function. The nitrogen is potentially stereogenic and as mentioned in section 1.7 inversion is usually negligible in such small rings with electronegative groups present.

When substituted by an electron-withdrawing substituent on nitrogen, oxaziridines become quite powerful oxidising agents, capable of epoxidising alkenes. Furthermore, when incorporated into an asymmetric framework oxaziridines such as (**69**) are useful reagents for the synthesis of enantiomerically enriched epoxides.

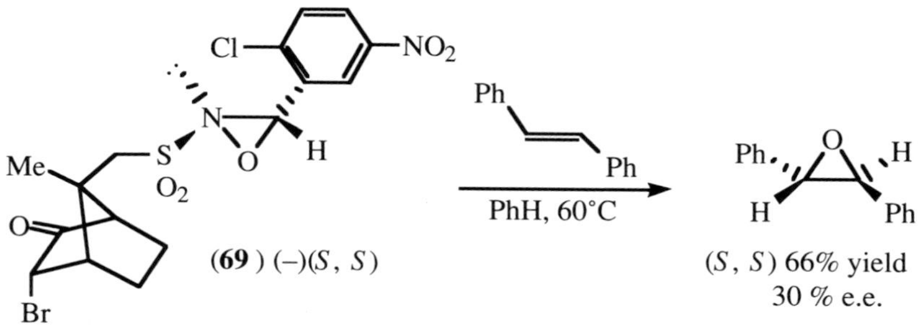

(**69**) (–)(*S, S*) → (*S, S*) 66% yield, 30 % e.e.

(PhH, 60°C; styrene Ph–CH=CH–Ph → epoxide)

Use of the (+) (*R, R*)-diastereomer of (**69**) at the oxaziridine (the rest of the molecule being unchanged) gives the (*R, R*) enantiomer of the epoxide in a similar e.e. of 35%, suggesting that it is the asymmetry at the oxaziridine which is critical, and not that of the pendant camphoryl group.

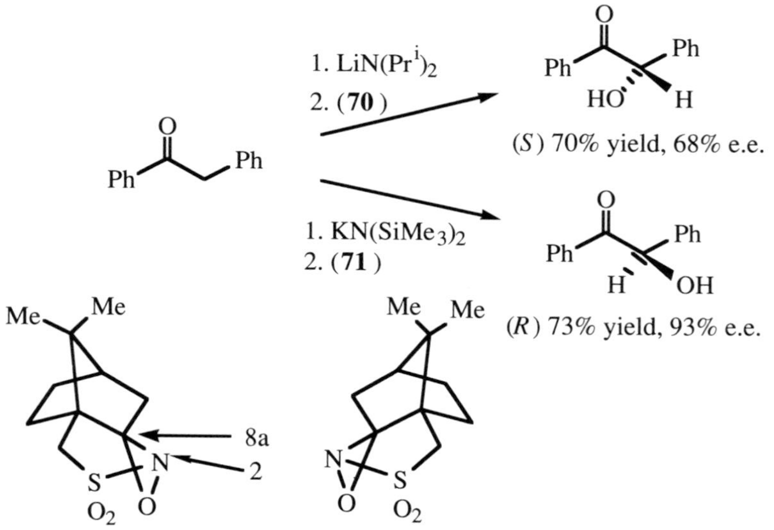

Ph–CO–CH₂–Ph
1. LiN(Pr^i)₂
2. (**70**)
→ Ph–CO–CH(OH)–Ph (*S*) 70% yield, 68% e.e.

1. KN(SiMe₃)₂
2. (**71**)
→ Ph–CO–CH(OH)–Ph (*R*) 73% yield, 93% e.e.

(**70**) (+)-(2*S*, 8a*R*)   (**71**) (–)-(2*R*, 8a*S*)

The sulphonyloxaziridines (**70**) and (**71**) derived from (+)- and (−)-camphor-sulphonic acid are very useful for the asymmetric synthesis of α-hydroxycarbonyl compounds, which are important and widely distributed. The reaction is performed simply by treating the prochiral enolate with the chiral oxidant, as in the example shown for preparation of either enantiomer of benzoin.

The reaction also works well with ester and amide enolates as shown below:

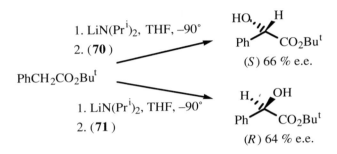

The enantiomeric excess is very sensitive to the reaction conditions. In some cases addition of $(Me_2N)_3PO$ (HMPA) improves the e.e., and in others it reduces it. Similarly, sometimes the potassium enolate gives better results than the lithium enolate.

6.4 Asymmetric reduction, double bond isomerisation and hydroboration

6.4.1 Catalytic hydrogenation with chiral transition metal complexes[27]

Only a brief overview can be given of the most important methods in this domain, since this is certainly one of the most widely exploited means of direct asymmetric synthesis, with many industrial applications.

Catalytic reduction with transition metal complexes lends itself well to asymmetric synthesis because of the rigid structure of these (generally) octahedral complexes and the resulting high degree of order in the transition state. The most widely used metals have been rhodium and ruthenium complexed to a huge range of chiral ligands, generally bisphosphines. A few of the more important and commercially available ones are shown below.[28]

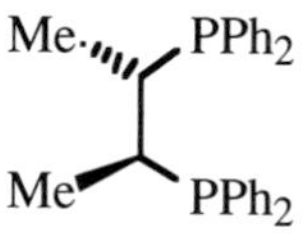

CHIRAPHOS

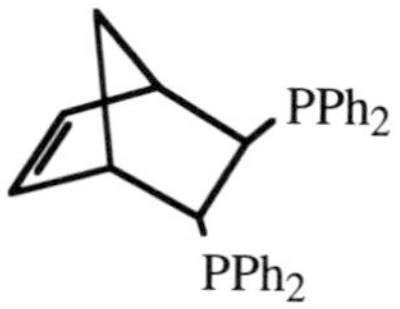

NORPHOS

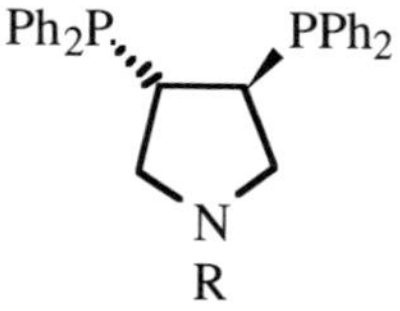

DEGPHOS

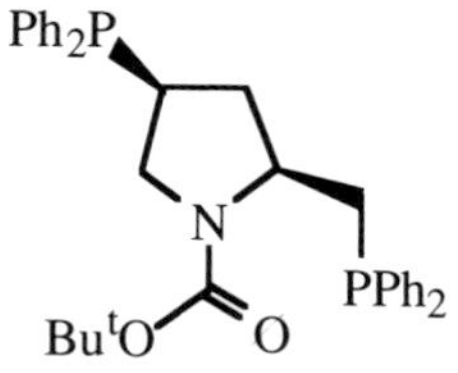

BPPM

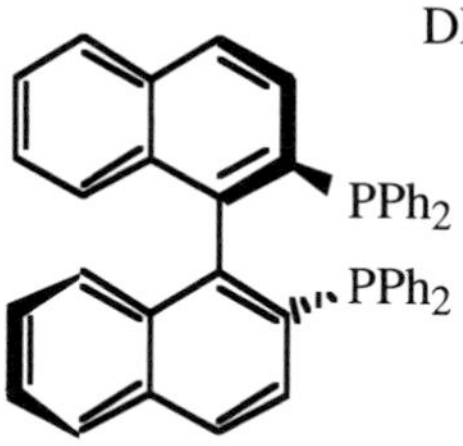

BINAP

Of these ligands, Noyori's BINAP is probably the most effective, sometimes giving products of almost 100% enantiomeric purity.[29] Accordingly, the following examples of asymmetric hydrogenation are largely based on this ligand, although other ligands often give excellent results. In any case, the basic principles are the same. The chiral catalysts are prepared from ruthenium(II) or rhodium(II) acetate and (R)- or (S)-BINAP.

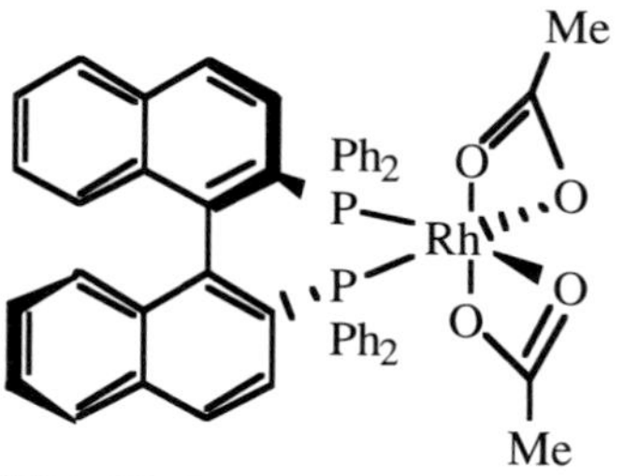

(72) (R)-BINAP-Rh(OAc)$_2$

Δ

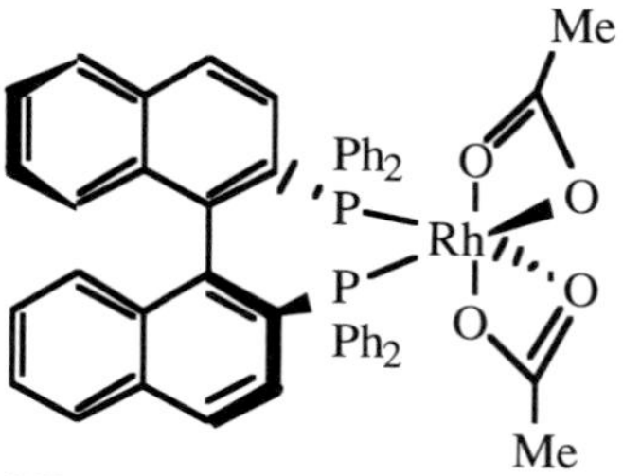

(73) (S)-BINAP-Rh(OAc)$_2$

Λ

THIRD- AND FOURTH-GENERATION METHODS

The complexes (**72**) and (**73**) are octahedral and, as described in section 6.1.2, can exist in two enantiomeric forms, Δ and Λ. In this case, the *R*-BINAP forms *exclusively* the *R*-Δ diastereomer, and the *S*-BINAP the *S*-Λ. A catalytic quantity (1 mol % or even less) promotes the asymmetric hydrogenation of a wide variety of alkenes bearing nearby polar groups, such as amides, alcohols, esters, etc., which are necessary for high e.e.s. An important application of this is in the synthesis of amino acids by asymmetric reduction of amidoacrylates (**74**).

This reaction is the key step in the industrial synthesis of (L)-DOPA, the anti-Parkinson's disease agent. In this case the hydrogenation is controlled by DIPAMP, a ligand chiral at phosphorus whose preparation was described in section 1.7.

170　ASYMMETRIC SYNTHESIS

The method has also proved to be applicable to isoquinoline alkaloid synthesis as illustrated by the example of the two enantiomers of tetrahydropapaverine (**75**) shown below. It is instructive to compare this fourth-generation approach with the second-generation method using chiral formamidines (section 5.3.4).

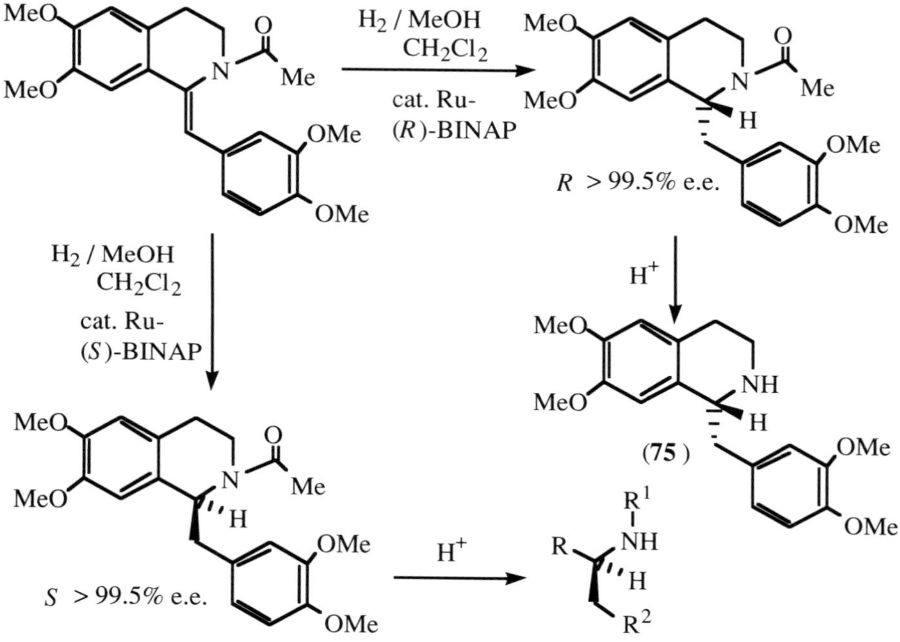

Prochiral allylic and homoallylic alcohols are reduced in very high optical yield with BINAP-Ru complexes. The stereochemical outcome depends not only on the chirality of the catalyst, but also on the geometry of the double bond, as evidenced by the synthesis of (*R*)- or (*S*)-citronellol (**76**) from nerol or geraniol. Note that the terminal double bond is not reduced, due probably to the absence of a nearby internal ligand.

Acrylic acids are also excellent substrates for this asymmetric reduction, as demonstrated by the synthesis of the anti-inflammatory drug (*S*)-naproxen (**77**).

The mechanism of these remarkable reductions is believed to be as follows: in solution the complex is in equilibrium with a four-coordinate square-planar form (**78**). The double bond substrate coordinates to this species largely on one enantiotopic face, as a result of the extremely asymmetric environment of metal nucleus. (Note the importance of the polar group as a ligand.) Hydrogen undergoes oxidative addition to the complex (**79**) at this point, and this is followed by hydrogen transfer from the metal to the alkene, resulting in a σ-bonded complex. The catalytic cycle is completed by reductive elimination of the reduced alkene from (**80**). The overall effect is that both hydrogens are delivered to the complexed face of the alkene, which explains why alkene geometry is critical.

It might be assumed from this that the enantioselectivity results from the chiral phosphine favouring the adduct (**79**) as opposed to its diastereomer which has the substrate bound the other way round. In

THIRD- AND FOURTH-GENERATION METHODS 173

fact there is now good evidence that it is *the less favoured diastereomer* (**79**) from which the observed product actually arises, because it is much more reactive.

Ketones bearing polar groups one to three carbons away can also be reduced with very high enantioselectivity to the secondary alcohol with ruthenium-BINAP complexes.

$$\underset{R}{\overset{HO\quad H}{\diagup}}\!\!\diagdown\!(C)_n\!-\!X \;\xleftarrow[\text{(R)-BINAP}]{H_2,\;\text{cat. Ru-}}\; R\!-\!\overset{O}{\underset{\|}{C}}\!-\!(C)_n\!-\!X \;\xrightarrow[\text{(S)-BINAP}]{H_2,\;\text{cat. Ru-}}\; \underset{R}{\overset{HO\quad H}{\diagup}}\!\!\diagdown\!(C)_n\!-\!X$$

n= 1-3, X= heteroatom

The method allows an extraordinarily easy and efficient synthesis of (*R*)- or (*S*)-3-hydroxyalkanoate esters, which are valuable synthetic intermediates.

$$\underset{\substack{R\\ >99\%\text{ e.e.}}}{\overset{H\quad OH\quad O}{\diagup}}\!\!\diagdown\!\text{OMe} \;\xleftarrow[\text{(R)-BINAP}]{\text{20 atm. }H_2\;\text{MeOH, cat. Ru-}}\; R\!-\!\overset{O\;\;O}{\underset{\|\;\;\|}{C}}\!-\!\text{OMe} \;\xrightarrow[\text{(S)-BINAP}]{\text{20 atm. }H_2\;\text{MeOH, cat. Ru-}}\; \underset{\substack{R\\ 96\%\text{ yield}\\ >99\%\text{ e.e.}}}{\overset{H\quad OH\quad O}{\diagup}}\!\!\diagdown\!\text{OMe}$$

There are two evident advantages over the enzymatic route (section 6.5.1). First, both enantiomers are available, whereas yeast reduction only gives the (*S*). Second, the reaction is far more convenient experimentally, as it does not involve large volumes of aqueous solvent, with attendant isolation problems.

6.4.2 Asymmetric double bond isomerisation[30]

Allylic amines are isomerised by transition metal catalysts to enamines. The isomerisation can be rendered asymmetric with rhodium-BINAP complexes.

174 ASYMMETRIC SYNTHESIS

[Scheme: Allylamines with Me, R, NEt$_2$ substituents undergoing H$_2$/Rh-(S)-BINAP or Rh-(R)-BINAP catalyzed isomerization to give (S)- or (R)-enamines. R = alkyl, aryl, etc.]

As can be seen from the example above, the enantiomer obtained again depends on the geometry of the starting alkene as well as the handedness of the catalyst. The mechanism is not fundamentally different from that of hydrogenation, except that the hydrogen is derived this time from the substrate rather than from hydrogen gas. For a remarkable industrial application of this asymmetric 1,3-hydrogen shift, see section 7.4.3.

6.4.3 Asymmetric hydroboration of alkenes[31]

Boron hydrides bearing chiral groups are valuable reagents for the transformation of prochiral Z-alkenes into enantiomerically enriched alcohols and many other compounds. The most widely used chiral borane is diisopinocampheylborane (Ipc$_2$BH) which is commercially available in both enantiomeric forms by the hydroboration of (+)-α-pinene and its enantiomer. Originally, the chiral boranes were prepared directly from commercial (+)- or (–)-α-pinene, which has an optical purity of only c. 93%. Subsequently, it was discovered that by equilibrating the Ipc$_2$BH with a 15% excess of α-pinene at 0°, the major enantiomer crystallised with an optical purity of 99%, leading to a corresponding improvement in the e.e. of the products.

THIRD- AND FOURTH-GENERATION METHODS

A simple example of the use of these chiral reagents is provided by the chiral hydroboration of Z-but-2-ene. Note that cleavage with iodine results in inversion at the stereogenic centre, unlike the other methods, which give retention.

The chiral hydroboration works particularly well with cyclic alkenes. It is a key step in an elegant asymmetric synthesis of loganin (**81**) and can be seen in action again in section 7.3.1.

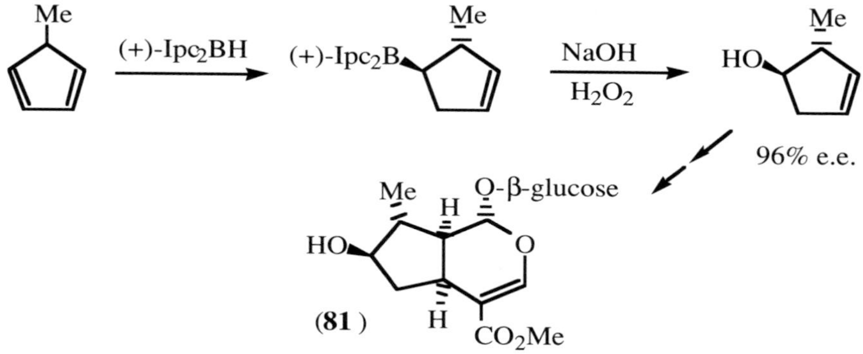

[Scheme showing cyclopentadiene + (+)-Ipc₂BH → (+)-Ipc₂B-cyclopentenyl → NaOH/H₂O₂ → HO-cyclopentenyl-Me, 96% e.e., leading to structure (81) with Me, HO, H, O-β-glucose, O, CO₂Me]

As noted above, only cyclic and Z-alkenes work well with Ipc$_2$BH. To obtain good optical yields with E- or trisubstituted alkenes, it is necessary to use IpcBH$_2$ as in the example below.

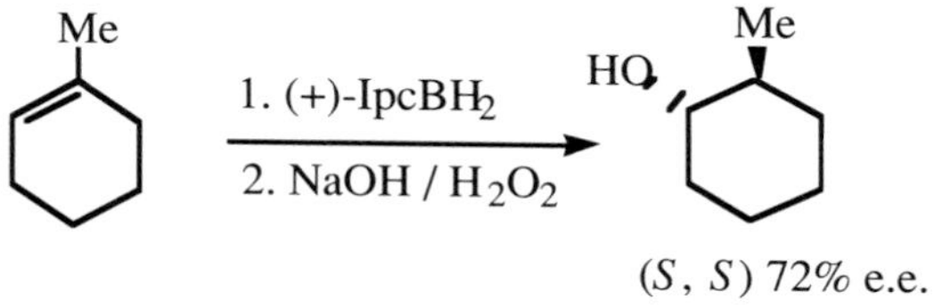

[Scheme: 1-methylcyclohexene + 1. (+)-IpcBH$_2$ 2. NaOH/H$_2$O$_2$ → trans-2-methylcyclohexanol, (S, S) 72% e.e.]

6.4.4 Asymmetric reduction using chiral boranes and borohydrides[32]

Alpine borane® is the delightful trivial name of the trialkylborane obtained from 9-borabicyclo[3.3.1]nonane and (+)- or (−)-α-pinene.

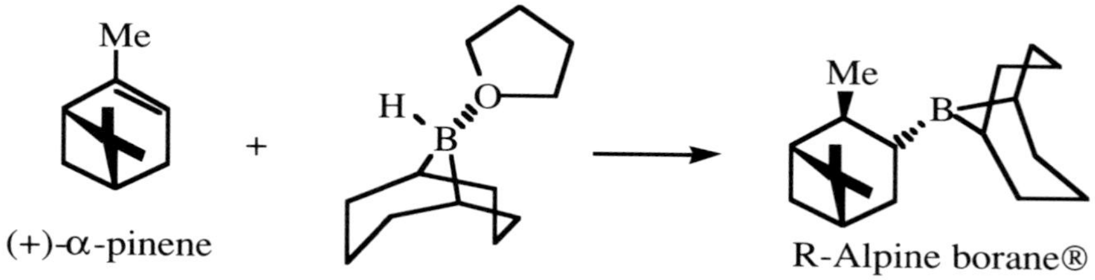

[Scheme: (+)-α-pinene + 9-BBN·THF → R-Alpine borane®]

The reagent reduces a variety of ketones directly to secondary alcohols with reasonable enantioselectivity, providing one substituent on the ketone is much bulkier than the other. It is particularly effective with acetylenic ketones.

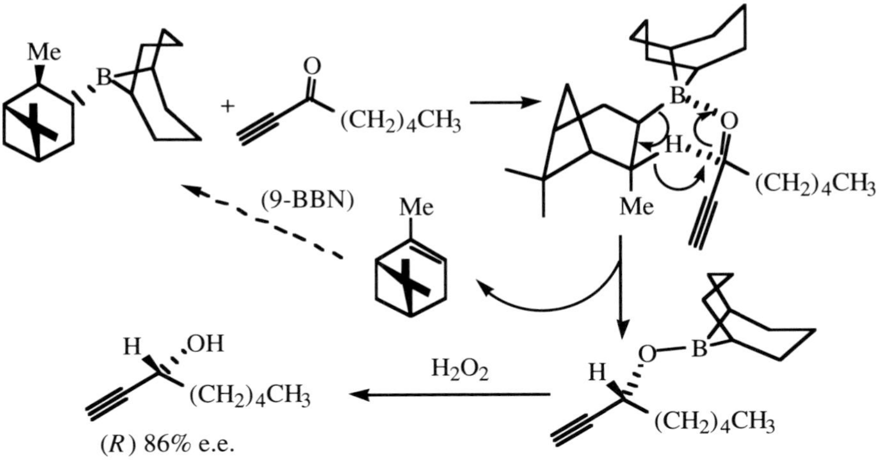

As can be seen from the diagram, the reaction probably proceeds by β-hydride transfer *via* a boat-shaped transition state in which the less bulky substituent buttresses the methyl group of the pinene, accounting for the high selectivities seen with ketones containing the slender ethynyl group. Note that the pinene is regenerated and can be reconverted to Alpine borane as shown. In this way (+)-α-pinene can be seen merely as an asymmetric carrier for the hydride of 9-BBN.

For more hindered ketones, the reagent Ipc_2BCl often gives excellent results. The mechanism is similar. For example:

Aryl ketones can be reduced with very high enantioselectivity using an 'ate' complex (**82**) prepared from 9-BBN and the saccharide derivative diisopropylidene-α-D-glucofuranose (DIPGF).

The reduction of phenyl *t*-butyl ketone is particularly impressive in view of the inertness of this substrate to other chiral reducing agents.

Perhaps the most general method of asymmetric ketone reduction is that recently developed by Corey.[33] The chiral reagents such as (**84**, R = H) are obtained by the reaction of diborane with an amino alcohol; the most effective chiral amino alcohol appears to be (**83**) derived from (*S*)-proline. The reduction of ketones occurs generally with good enantioselectivity (80–97% e.e.), and is actually *catalytic* in the chiral reagent, since reduction of ketones by diborane itself is very slow.

As an example, acetophenone (R_S = methyl, R_L = phenyl) is reduced with 1 equivalent of diborane and 2.5 mol % of the catalyst (**84**, R = H) to (*R*)-1-phenylethanol with an e.e. of 97% in only one minute at room temperature in THF.

More recently the β-alkyl compounds (**84**) formed by reaction of (**83**) with boronic acids have proved to be even more effective and stable catalysts. The mechanism for the catalytic cycle of the reduction is shown for (**84**, R = Me). Corey has coined the term 'chemzyme' for this type of enzyme-like catalyst.

Catechol borane can also be used as the reducing agent with (**84**, R = Bu^n) as catalyst to give alcohols in excellent yield and >90% e.e. with predictable absolute configuration as shown.

6.4.5 Chirally modified LiAlH$_4$[34]

Though somewhat overshadowed nowadays by the more modern asymmetric reduction methods (chiral boranes, BINAP complexes, etc.), the reduction of ketones with lithium aluminium hydride modified by chiral ligands is still of use and interest.

The most effective chiral ligands out of the dozens tried are bidentate, i.e. those which can form a cyclic complex around the aluminium atom. The proline-based diamine (**85**) of Mukaiyama is a good example.

Although not proposed by the authors, it can be safely assumed that the reaction is proceeding through a six-membered cyclic transition state.

The reaction works well for aromatic ketones (e.e. 86–96%), but, as one might expect, less well for aliphatic ketones, in which the two substituents are of more equivalent steric bulk.

The second example in this subsection concerns the synthesis of an intermediate in a prostaglandin synthesis. The chiral additive in this case is the amino alcohol *N*-methylephedrine (**86**). See section 7.4.4 for an alternative enzymatic route to the product.

6.5 Enzymatic and microbial methods[35]

In recent years there has been a veritable renaissance in the field of enzymology, due in part to the realisation that many enzymes will accept 'unnatural' substrates which can be used to probe their active sites. For synthetic chemists, this 'promiscuousness' is of great interest and importance, since enzymes are fourth-generation asymmetric catalysts *par excellence*. As we shall see, an understanding of the enzyme mechanism though desirable, is often unnecessary, allowing one to use an enzyme like any other reagent. Sometimes indeed it is not even necessary to isolate the enzyme.

6.5.1 Enzymatic reduction

Most asymmetric reductions effected enzymatically have been of ketones.[36] The reactions have been conducted with whole cells (usually yeasts) as well as with isolated enzymes. In the latter case, of course, it is necessary to add at least an equivalent of the cofactor NADH or NADPH (nicotinamide adenine dinucleotide) as the actual reductant.

The dihydroxyacetone reductase of *Mucor javanicus* is NADPH-dependent, and accepts a wide variety of simple acyclic ketones. For 2-alkanones, hydride attack occurs largely from the *Si* face with high selectivity to give the (*R*)-product. However, butan-2-one was reduced to racemic butan-2-ol, indicating that methyl and ethyl are not differentiated by the enzyme.

$$\text{pentan-2-one} \xrightarrow[\text{NADPH}]{\text{DHA reductase}} (R)\text{-pentan-2-ol, 94\% e.e.}$$

The opposite enantiomer predominates in the reduction of simple ketones by baker's yeast (*Saccharomyces cerevisiae*).[37]

$$\text{acetophenone} \xrightarrow{\text{Yeast}} (S)\text{-1-phenylethanol, 89\% e.e.}$$

The synthetic value of *S. cerevisiae* is increased by the fact that its oxidoreductase(s) will accept haloketones as substrates, reducing them with reasonable enantioselectivity to the (*R*) enantiomer. (Note that the stereochemical outcome is the same, despite the apparent inversion due to the CIP rules.)

$$\text{PhCOCH}_2\text{X} \xrightarrow[X = Cl, Br]{\text{Yeast}} (R)\text{-PhCH(OH)CH}_2\text{X}, >80\% \text{ e.e.}$$

The most widely used substrates in reduction by yeast have been β-ketoesters and β-ketoacids, since the products are valuable chiral building blocks. The enantiomer obtained and the e.e. are critically dependent on the substituents.

THIRD- AND FOURTH-GENERATION METHODS

[Scheme: β-ketoester reduction by yeast giving (S)-hydroxy ester (left) and (R)-hydroxy ester (right)]

R = Et, X = Et	e.e. 40%	R = Me, X = OEt	e.e. 95%
R = nPr, X = OH	e.e. 100%	R = Et, X = OC_8H_{17}	e.e. 99%
R = $ClCH_2$, X = OEt	e.e. 55%	R = $ClCH_2$, X = NHPh	e.e. 97%
R = Ph, X = OEt	e.e. 100%		

Generally, the best enantioselectivities are obtained at low substrate concentrations. This is conveniently achieved by slow addition.

A complication arises if the β-dicarbonyl compound is substituted on the central carbon, since diastereomers could be formed. Nonetheless useful selectivity is often observed. The enzyme D-3-hydroxybutyrate dehydrogenase (HBDH) from *Rhodopseudomonas spheroides* in the presence of NADH reduces only the (S)-enantiomer of 2-ketocyclohexanecarboxylic acid,[38] whereas baker's yeast operates exclusively on the (R)-antipode.

[Scheme: tautomeric equilibrium of 2-ketocyclohexanecarboxylate between (S) and (R) forms via enol (R = H, Et); (S) reduced by HBDH to cis hydroxy ester 100% e.e.; (R) reduced by Yeast to hydroxy ester 100% e.e.]

Enzymatic reduction is not limited to β-ketoesters and acids: baker's yeast reduces a number of α-ketoacids with good chemical yield and enantioselectivity.

[Scheme: PhCOCO₂Me → Yeast → PhCH(OH)CO₂Me, 59% yield, 100% e.e.]

[Scheme: R-CO-CH₂CH₂-CO₂H → γ-butyrolactone (87) with R substituent]

R = n-alkyl, C₂–C₇

[Scheme: R-CO-CH₂CH₂CH₂-CO₂H → δ-valerolactone (88) with R substituent]

Similarly, γ- and δ-ketoacids are reduced enantioselectively by the bacterium *Sarcina lutea* and the mould *Cladosporium butyri* respectively. The products are isolated as the lactones (87) and (88).[39]

Of industrial importance in steroid synthesis, the asymmetric reduction of 2,2-disubstituted cyclopentane-1,3-diones such as (89) has been investigated in detail.[40]

[Scheme: (89) → *Saccharomyces uvarum* → (2R,3S) 76% yield → steroid product]

Note that this reduction is diastereoselective as well as enantioselective: only a small amount of the (2S, 3S) diastereomer is formed.

THIRD- AND FOURTH-GENERATION METHODS   185

Enzymic reduction is seen in action in total synthesis in section 7.4.4.

6.5.2 *Enantioselective microbial oxidations*

An enzymatic reaction of some synthetic importance is the enantioselective oxidation of *meso* diols by horse liver alcohol dehydrogenase HLADH.[41] Some representative examples follow:

[Structure: bicyclic diene diol] → HLADH, NAD / FMN → [bicyclic lactone]
64% yield, >97% e.e.

[Structure: bicyclic diol] → HLADH, NAD / FMN → [bicyclic lactone]
87% yield, >97% e.e.

[Structure: cis-cyclohexane diol] → HLADH, NAD / FMN → [bicyclic lactone]
76% yield, 100% e.e.

There are two interesting features to the reaction. First, HLADH accomplishes two oxidations, to the hemiacetal and thence to the lactone. Second, as in the natural system (a horse), the four hydrogens from the substrate end up on the flavin mononucleotide cofactor (FMN). The NAD is recycled, and is therefore only needed in 'catalytic' amount.

The enantiomerically pure lactones which result from HLADH oxidations can be transformed by standard chemistry into valuable chiral units such as (**90**).

The humble soil bacteria of *Pseudomonas* species are capable of effecting a variety of enantiospecific *cis*-dihydroxylations of alkenes and aromatics, some of which have no equivalent in ordinary organic chemistry.[42]

>95% e.e.

The product from Pseudomonas oxidation of toluene can be converted in only three steps to a prostaglandin intermediate (**91**).

Such arenediols (so-called) are also of interest as chiral intermediates in terpenoid synthesis as illustrated by the synthesis of both enantiomers of (**92**).

We can confidently expect to see further applications of *Pseudomonas* sp. oxidation in the future.

6.5.3 Esterases and lipases[43]

Enzymes which catalyse the hydrolysis of esters and lipids are often capable of high enantioselection. Pig liver esterase (PLE) has been widely used for the enantioselective synthesis of half-esters from diesters.[44] Methods based on such enzymes have the advantage over oxidases in that no cofactor is necessary.

As can be seen, it is easy to make either enantiomer of the lactone by appropriate choice of reducing agent. PLE gives good results for *cis*-cyclopropane- and *cis*-cyclobutanedicarboxylic esters, but only 17% e.e. for cyclopentane. Increasing the ring size to six restores the enantioselectivity, but in the opposite enantiomeric series, indicating that the diester is now binding in a different way.

The enzyme can also operate on diacetates, though with much poorer enantioselectivity. For such substrates lipases usually give better results.[45]

83% e.e.

Lipases, like all catalysts, are capable of catalysing the reverse reaction, namely acylation of alcohols. This is the basis of an interesting kinetic resolution.[46] The lipase from *Pseudomonas* sp. is treated with a racemic substituted cyclohexanol and vinyl acetate. The latter reagent acylates the active site of the enzyme, giving the acyl enzyme, which then specifically transfers the acetyl group to the (R)-enantiomer of the cyclohexanol. It is then a simple matter to separate the (S)-alcohol from the (R)-acetate.

(R= Ph, OPh, OCH$_2$Ph)

50- >98% e.e.   ≥98% e.e.

The use of an esterase in total synthesis is described in section 7.4.5.

References

1. J.D. Morrison and J.W. Scott, eds., *Asymmetric Synthesis*, Vol. 1–5, Academic Press, Orlando, 1983–5.
2. D.L. Hughes, U.H. Dolling, K.M. Ryan, E.F. Schoenmaldt and E.J.J. Grabowski, *J. Org. Chem.*, 1987, **52**, 4745.
3. H. Wynberg, *Top. Stereochem.*, 1986, **16**, 87.

4. R.S.E. Conn, A.V. Lovell, S. Karady and L.M. Weinstock, *J. Org. Chem.*, 1986, **51**, 4710.
5. K. Hermann and H. Wynberg, *J. Org. Chem.*, 1979, **44**, 2238.
6. H. Brunner and B. Hammer, *Angew. Chem., Int. Ed. Engl.*, 1984, **23**, 312.
7. K. Suzuki, A. Uregama and T. Mukaiyama, *Bull. Chem. Soc. Jpn.*, 1982, **55**, 3277.
8. R. Noyori and M. Kitamura, *Angew. Chem., Int. Ed. Engl.*, 1991, **30**, 49.
9. K. Soai, S. Numa and M. Watanabe, *J. Org. Chem.*, 1988, **53**, 927.
10. L.A. Paquette, in *Asymmetric Synthesis*, Volume 3, J.D. Morrison, ed., Academic Press, Orlando, 1984, Chapter 7.
11. Y. Ichikauna, A. Narita, A. Shiozauna, Y. Hayashi and K. Narasaka, *J. Chem. Soc., Chem. Commun.*, 1989, 1919.
12. M. Maruoka, T. Itoh, T. Shirasaka and H. Yamamoto, *J. Am. Chem. Soc.*, 1988, **110**, 310.
13. D.S. Matteson, *Synthesis*, 1986, 973.
14. S. Hanessian, D. Delorme, S. Beaudoin and Y. Leblanc, *J. Am. Chem. Soc.*, 1984, **106**, 5754.
15. L. Fuji, M. Node, S. Terada, M. Murata, H. Nagasauna, T. Taga and K. Machida, *J. Am. Chem. Soc.*, 1985, **107**, 6404.
16. L. Duhamel, P. Duhamel, J.C. Launay and J.C. Plaquevent, *Bull. Soc. Chim. Fr. II*, 1984, 421.
17. P.J. Cox and N.S. Simpkins, *Tetrahedron: Asymmetry*, 1991, **2**, 1.
18. Z.G. Hajos and D.R. Parrish, *J. Org. Chem.*, 1974, **39**, 1615.
19. J. Gutziniller, P. Buchschacher and A. Fürst, *Synthesis*, 1977, 167.
20. Synthetic applications: B.E. Rossiter, in *Asymmetric Synthesis*, Volume 5, J.D. Morrison, ed., Academic Press, Orlando, 1985, Chapter 7; A. Pfenninger, *Synthesis*, 1986, 89.
    Mechanism: M.G. Finn and K.B. Sharpless, in *Asymmetric Synthesis*, Volume 5, J.D. Morrison, ed., Academic Press, Orlando, 1985, Chapter 8.
21. H.B. Kagan and F. Rebiere, *Synlett.*, 1990, 643.
22. S.G. Hentges and K.B. Sharpless, *J. Am. Chem. Soc.*, 1980, **103**, 4263.
23. E.N. Jacobson, I. Markó, W.S. Mungall, G. Schröder and K.B. Sharpless, *J. Am. Chem. Soc.*, 1988, **110**, 1968.
24. B.B. Lohray, T.H. Kalantar, B.M. Kim, C.Y. Park, T. Shibata, J.S.M. Wai and K.B. Sharpless, *Tetrahedron Lett*, 1989, **30**, 2041.
25. J.S.M. Wai, I. Markó, J.S. Svendsen, M.G. Finn, E.N. Jacobsen and K.B. Sharpless, *J. Am. Chem. Soc.*, 1989, **111**, 1123.
26. F.A. Davis and A.C. Sheppard, *Tetrahedron*, 1989, **45**, 5703; F.A. Davis, in *Asymmetric Synthesis*, Volume 4, J.D. Morrison and J.W. Scott, eds., Academic Press, Orlando, 1984, Chapter 4.
27. Mechanism: J. Halpern, in *Asymmetric Synthesis*, Volume 5, J.D. Morrison, ed., Academic Press, Orlando, 1985, Chapter 2.
    Applications: K.E. Koenig, in *Asymmetric Synthesis*, Volume 5, J.D. Morrison, ed., Academic Press, Orlando, 1985, Chapter 3.
28. H.B. Kagan, in *Asymmetric Synthesis*, Volume 5, J.D. Morrison, ed., Academic Press, Orlando, 1985, Chapter 1.
29. R. Noyori and T. Takaya, *Acc. Chem. Res.*, 1990, **23**, 345.
30. S. Otsuka and K. Tani, *Synthesis*, 1991, 665; see also S. Otsuka and K. Tani in *Asymmetric Synthesis*, Volume 5, J.D. Morrison, ed., Academic Press, Orlando, 1985, Chapter 6.
31. H.C. Brown and P.K. Jadhav, in *Asymmetric Synthesis,* Volume 2, J.D. Morrison, ed., Academic Press, Orlando, 1983, Chapter 1; H.C. Brown and B. Singaram, *Acc. Chem. Res.*, 1988, **21**, 287; M. Srebnik and P.V. Ramachandran, *Aldrichimica Acta*, 1987, **20**, 9.

32. M.M. Midland, in *Asymmetric Synthesis*, Volume 2, J.D. Morrison, ed., Academic Press, Orlando, 1983, Chapter 2; M.M. Midland, *Chem. Rev.*, 1989, **89**, 1553; H.C. Brown and P.V. Ramachandran, *Acc. Chem. Res.*, 1992, **25**, 16.
33. E.J. Corey and R.K. Bakshi, *Tetrahedron Lett.*, 1990, **31**, 611, and previous work cited therein.
34. E.R. Grandbois, S.I. Howard and J.D. Morrison, in *Asymmetric Synthesis*, Volume 2, J.D. Morrison, ed., Academic Press, Orlando, 1983, Chapter 3.
35. J.B. Jones, in *Asymmetric Synthesis*, Volume 5, J.D. Morrison, ed., Academic Press, Orlando, 1985, Chapter 9; H.G. Davies, R.H. Green, D.R. Kelly and S.M. Roberts, *Biotransformations in Preparative Organic Chemistry*, Academic Press, 1989.
36. C.J. Sih and C.S. Chen, *Angew. Chem., Int. Ed. Engl.*, 1984, **23**, 570.
37. S. Servi, *Synthesis*, 1990, 1; R. Csuk and B.I. Glänzer, *Chem. Rev.*, 1991, **91**, 49.
38. S.A. Brenner and T.H. Morton, *J. Am. Chem. Soc.*, 1981, **103**, 991.
39. G.T. Muys, B. Van der Ven and A.P. De Jorge, *Appl. Microbiol.*, 1963, **11**, 389.
40. K. Kieslich, in *Microbial Transformations of Non-Steroid Compounds*, Thieme, Stuttgart, 1976, 29, and references therein.
41. K.P. Lok, I.J. Jakovac and J.B. Jones, *J. Am. Chem. Soc.*, 1985, **107**, 2521; A.J. Bridges, P.S. Raman, G.S.Y. Ng, and J.B. Jones, *J. Am. Chem. Soc.*, 1984, **106**, 1461.
42. T. Hudlicky, H. Luna, G. Barbieri and L.D. Kunart, *J. Am. Chem. Soc.*, 1988, **110**, 4735.
43. W. Boland, C. Frössl and M. Lorenz, *Synthesis*, 1991, 1049.
44. L.-M. Zhu and M.C. Tedford, *Tetrahedron*, 1990, **46**, 6587.
45. C.S. Chen and C.J. Sih, *Angew. Chem., Int. Ed. Engl.*, 1989, **28**, 695.
46. K. Laumen, R. Seemayer and M.P. Schneider, *J. Chem. Soc., Chem. Commun.*, 1990, 49.

7 Asymmetric total synthesis
S.N. KILENYI

These days organic chemists are no longer content merely to synthesise target molecules in racemic form or as mixtures of stereoisomers. Instead it has become *de rigeur* to prepare compounds in enantiomerically pure or at least enantiomerically enriched form. This movement, which has been given additional impetus by the statutory requirement in some countries for new pharmaceutical products to be licensed as pure stereoisomers, has led to a veritable explosion of activity in the area of asymmetric synthesis. Because of this, only a representative selection of the hundreds of asymmetric total syntheses can be described in this chapter.

The examples are organised according to the classification of first- to fourth-generation methods introduced in chapter 4. Since the asymmetric steps are invariably accompanied by a large number of steps in which no new stereogenic unit is formed, the former are highlighted, with a number corresponding to the type of method involved, and it is hoped that this will assist the reader in identifying the key processes in each synthesis.

7.1 First-generation methods

7.1.1 *(R)-(+)-Frontalin* (Magnus and Roy[1])

The title compound is one of the sex pheromones of the Dutch Elm beetle *Scolytus scolytus*. For maximum biological activity it is essential for the pheromone to have a high degree of enantiomeric purity. This is achieved with a starting material from the pool of naturally occurring chiral compounds, (3R)-(−)-linalool, and in the process illustrates a new homologation procedure.

The synthesis commences by O-protection and selective ozonolysis of the more electron-rich double bond, giving aldehyde (**1**). This reacts in a fashion reminiscent of the Darzens condensation with the carbanionic reagent developed previously by Magnus, affording epoxysilane (**2**) as a mixture of all four possible diastereomers. After cleavage of the remaining double bond and complete reduction of the ozonide with borohydride, the product (**3**) is desilylated and rearranged with Lewis acid giving initially the ketone (**4**). Under the reaction conditions this undergoes internal ketalisation to the final product. It is only in this final step that a new stereocentre is formed making it a true asymmetric synthesis, as is clear when we compare the structure of (**4**) with the starting linalool. The preceeding steps serve only to convert the available chiral precursor into a form suitable to undergo the first-generation asymmetric process of ketalisation. This is completely diastereoselective since the alternative mode of cyclisation is geometrically impossible. The natural and synthetic samples had virtually the same optical rotation, indicating that little or none of the unnatural enantiomer is present.

7.1.2 (−)-Prostaglandin E_1 (Stork and Takahashi[2])

The prostaglandins comprise a class of hormones widely distributed in mammalian tissue with actions ranging from inflammation and mediation of pain to platelet aggregation and sexual function. This synthesis of one of the most common prostaglandins, PGE_1, illustrates how an implausible chiral material like a carbohydrate can be used to make a totally different target molecule.

The synthesis starts with mannitol, which is a symmetrical but chiral compound (note the C_2-axis of symmetry). The bis-ketal is oxidatively cleaved to give two moles of a widely used chiral building block, the ketal of (+)-glyceraldehyde (**5**). Reaction of (**5**) with the anion of methyl oleate gives hydroxyester (**6**), hydroxyl protection

194 ASYMMETRIC SYNTHESIS

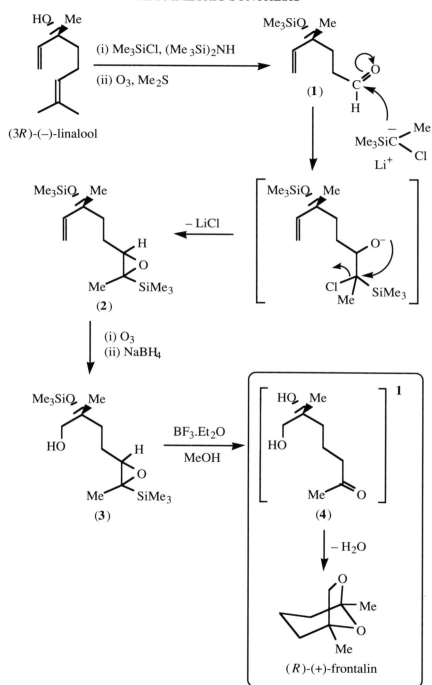

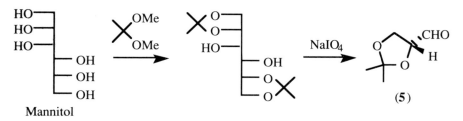

and acid treatment of which furnishes, by spontaneous lactonisation, lactone (**7**). This in turn is converted *via* (**8**) to the protected cyanohydrin (**9**). Stork had previously developed such reagents as acyl anion equivalents. Treatment with base results in intramolecular displacement of tosylate to give the cyclopentane (**10**). Oxidative cleavage of the double bond in the side-chain and selective deprotection of the more acid-labile ethoxyethyl groups gives (**11**). Mild base treatment results in collapse of the cyanohydrin and beta-elimination of the protected hydroxyl group to afford the simple hydroxycyclopentenone (**12**), which still contains only the one stereogenic centre from the glyceraldehyde. Because of this feature the stereoselectivity of all the stages from (**5**) to (**12**) is unimportant and only the key stereocentre must remain intact. It should again be noted that although the sequence from mannitol to (**12**) involves the formation of several new stereogenic units they are all subsequently destroyed and it is only the final step (**12**) → (**14**) which is a first-generation process. After hydroxyl protection (**12**) is reacted with three equivalents of the *racemic* cuprate (**13**). Remarkably, diastereo-differentiation occurs, which is to say that only *one* enantiomer of the cuprate reacts with the enantiomerically pure enone. Furthermore, attack of the cuprate occurs from the face *anti* to the bulky THP ether, as one would expect from steric considerations. Similarly, the resulting copper enolate species protonates on work-up from the less-hindered face *anti* to the newly-installed vinyl side-chain, giving (**14**), which has the wrong configuration at position 2. However, (**14**) can be converted to (−)-PGE$_1$ by known procedures. Comparison of the much shorter fourth-generation synthesis of (−)-PGE$_1$ by Sih (section 7.4.4) is instructive.

7.1.3 The Merck synthesis of (+)-thienamycin (Reider and Grabowski[3])

Thienamycin is an important antibiotic used clinically against a wide variety of pathogenic bacteria, including strains resistant to penicillins because of their production of the enzyme β-lactamase.

ASYMMETRIC TOTAL SYNTHESIS

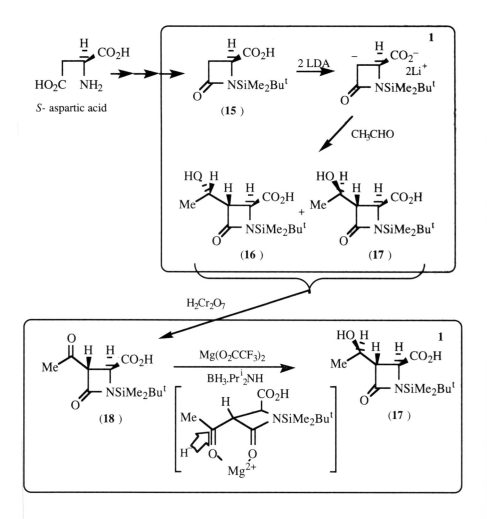

An unusual feature of the asymmetric synthesis described here is the temporary destruction of the original stereogenic centre.

The naturally occurring enantiomerically pure starting material is (S)-aspartic acid, which is converted by standard means into the β-lactam (**15**). This forms a dianion on treatment with LDA which reacts with acetaldehyde to give a mixture of aldol products (**16**) and (**17**). Note that attack of the acetaldehyde has come exclusively from the face opposite the carboxylate, thereby establishing the second stereogenic centre with both the correct relative (*anti*) and absolute (*R*) configurations. The mixture of alcohols is oxidised to the ketolactam (**18**) which, in the presence of magnesium ion, undergoes a stereospecific reduction to (**17**). On treatment with lead tetra-acetate, (**17**) undergoes oxidative decarboxylation to (**20**), probably *via* iminium ion (**19**). Note that this intermediate lacks the original stereogenic centre of the aspartic acid precursor. Fortunately the stereochemical information is recorded in the adjacent centre, which directs the attack of acetate ion to the *anti* face. The same arguments apply to the reaction of (**20**) with silyl enol ether (**21**). The product (**22**), which contains all three stereogenic centres of thienamycin, is cyclised by treatment with rhodium(II) acetate to (**23**), which is then transformed by standard chemistry into the antibiotic.

7.2 Second-generation methods

7.2.1 *Prelog–Djerassi lactone* (Evans and Bartroli[4])

The Prelog–Djerassi lactone is a degradation product of the macrolide antibiotics methymycin and narbomycin. Evans uses it here to illustrate his asymmetric enolate methodology (see section 5.3.2). Of the many syntheses of this popular target molecule, this is one of the best in terms of diastereo- and enantioselectivity.

The chiral auxiliary is the oxazolidinone (**24**) derived from (1*S*, 2*R*)-norephedrine. Acylation with propionyl chloride gives (**25**) and this is deprotonated to afford exclusively the internally chelated *Z*-enolate, which reacts with methallyl iodide from the face opposite the methyl and phenyl groups of the auxiliary. The product (**26**), a 97:3 mixture of diastereomers, is purified to a ratio of better than 500:1. Reductive removal of the auxiliary and careful oxidation of the primary alcohol under non-racemising conditions affords the chiral (*S*)-aldehyde (**27**). This in turn is reacted with the *boron* enolate of (**25**), which furnishes with remarkable selectivity the *u* aldol product (**28**). The reason for the choice of boron rather than lithium is to invert the facial selectivity of the reaction—the enolate is no longer constrained to be planar by internal chelation and rotates in order to place the bulky dibutyl borinyl group on the opposite side to the methyl and phenyl:

The aldehyde now attacks from the less-hindered side of the enolate, again opposite the substituents of the auxiliary, *via* the Zimmerman–Traxler six-membered cyclic transition state:

Note that the natural tendency of the aldehyde to undergo nucleophilic attack according to Cram's rule is overridden. The selectivity of this reaction surprised even the authors, who found the product was a 400:1 mixture of *u* and *l* diastereomers at the two newly created stereogenic centres, with an overall asymmetric induction with respect to the chiral auxiliary of no less than 660:1.

After hydroxyl protection, the remaining stereogenic centre is installed by hydroboration with the bulky *t*-hexyl borane, which proceeds with a modest but useful 1,3-diastereoselectivity of 85:15. Finally, the terminal hydroxyl group is oxidised to the acid, and the chiral auxiliary removed with base. The final product was crystallised to a diastereomeric and enantiomeric purity of better than 99.9%.

7.2.2 (R)-Muscone (Nelson and Mash[5])

Muscone is one of the components of the anal gland secretion of the civet cat, and despite its provenance, is much valued in perfumery for its 'musky' odour. The synthesis here illustrates an unusual concept in asymmetric synthesis: that of a chiral protecting group as an auxiliary.

The chiral auxiliary (**29**) was synthesised by standard means from unnatural (2S, 3S)-tartaric acid.

(S,S)-tartaric acid → (**29**)

Cheap and readily available cyclopentadecanone is alpha-brominated to (**30**) and ketalised with (**29**), and the resulting α-bromoketal (**31**) subjected to base induced elimination of HBr, giving (**32**). The key asymmetric step is now at hand. The Simmons-Smith cyclopropanation proceeds with very high facial selectivity, giving almost entirely the diastereomer of (**33**) shown. The result is rationalised on the basis of zinc chelation by the oxygens of the ketal. The synthesis is completed by acid hydrolysis of the ketal (the chiral diol may be recovered for reuse), reductive opening of the cyclopropane ring under Birch conditions, and re-oxidation of the alcohol back to the ketone.

7.2.3 (−)-Podophyllotoxin (Meyers et al.[6])

One of the most important discoveries made by the natural product chemist Kupchan was the unusual lignan-derived podophyllotoxin, which has clinical applications as a powerful and selective antineoplastic. The problem of synthesising it asymmetrically is solved elegantly using the chiral oxazoline auxiliary (see section 5.4.1).

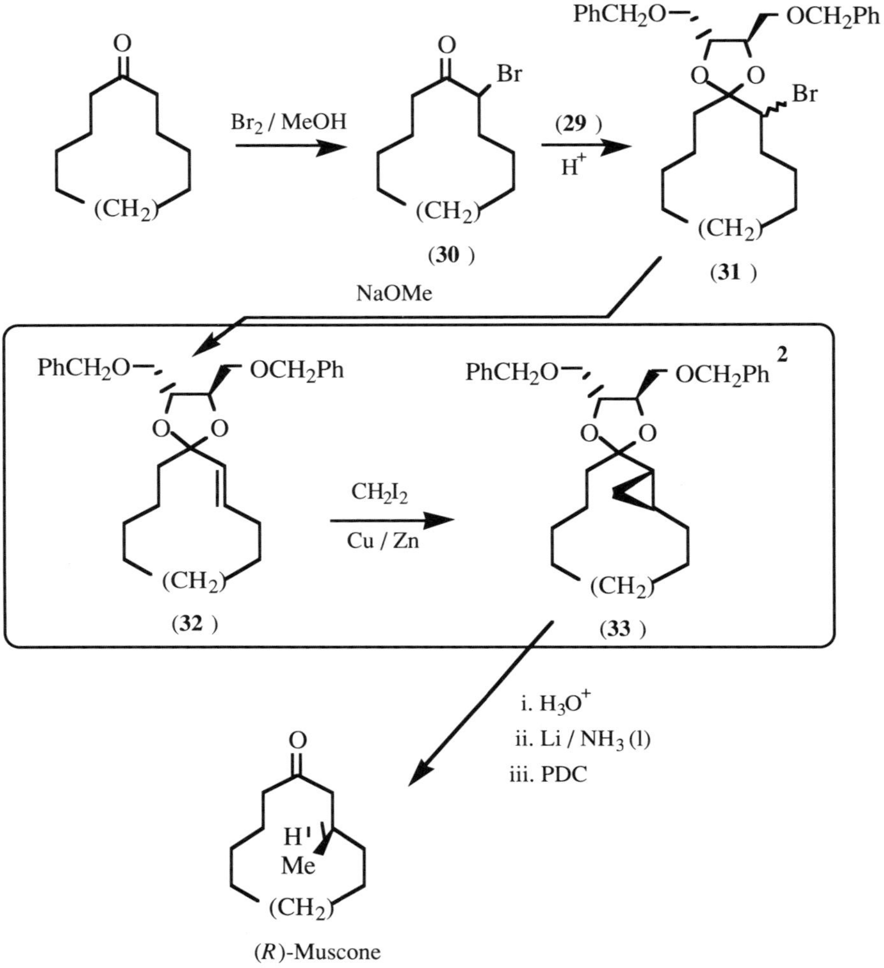

The starting material for this synthesis is the highly substituted achiral naphthalene (**34**), whose preparation is not given here. The ester group of (**34**) reacts with the chiral amino alcohol (**35**) derived from threonine to give the oxazoline (**36**), which reacts with aryllithium (**37**) diastereoselectively, affording dihydronaphthalene (**38**) as a 92:8 mixture of diastereomers. The alkoxy group on the auxiliary is presumed to deliver the nucleophile by chelation to the metal as shown opposite.

ASYMMETRIC TOTAL SYNTHESIS 205

Removal of the auxiliary and the allyl group gives lactone (**39**). On hydrolysis the double bond migrates back into conjugation with the ring, affording (**40**) after silylation and esterification. The double bond is then transformed *via* the bromohydrin to ketone (**41**), which is unfortunately epimeric at two of its three stereogenic centres with the final molecule. In order to invert these two centres a rather elaborate sequence involving hydroxymethylation [to give (**42**)], and a retro-aldol reaction to (**43**) is necessary. The ketone group of (**43**) is reduced stereospecifically to the desired configuration and to complete the synthesis the centre adjacent to the lactone carbonyl is inverted by protonation of the enolate. Separation of the two epimers (**44**) and (**45**) and desilylation gives (−)-podophyllotoxin with a high enantiomeric excess of 93–94%.

7.2.4 (−)-Steganone (Meyers et al. [7])

One of the most interesting compounds characterised by Kupchan in the 1970s is the lignan (−)-steganone, isolated from *Steganotaenia araliacea*. Its interest derives not only from its antileukaemic properties, but also from its axial chirality, which is rare among natural products.

This distinguishes steganone from all the other examples in this chapter as it is the only one which contains a stereogenic axis in addition to stereogenic centres. The key feature of the following

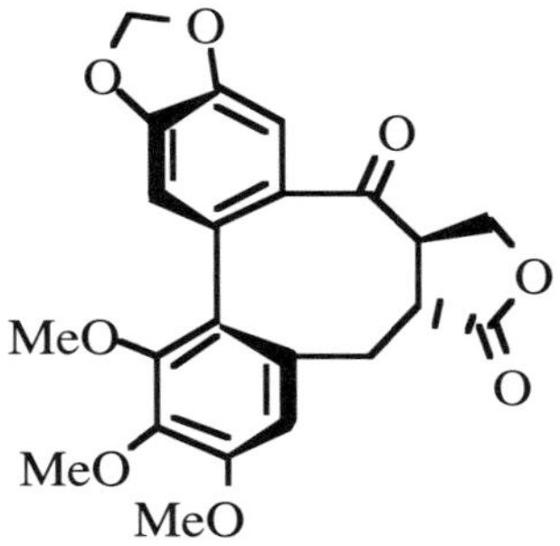

(−)-Steganone

synthesis, which differentiates it from the previous asymmetric syntheses of this target, is that the chiral biaryl moiety is formed first and then used to direct formation of the two stereogenic centres.

The two halves of the biaryl are constructed by standard chemistry. The upper half (**47**) is derived from piperonal by nuclear bromination to (**46**), followed by aldehyde protection as the acetal. As will be seen below, the state of hybridisation of this group is critical to the success of the synthesis.

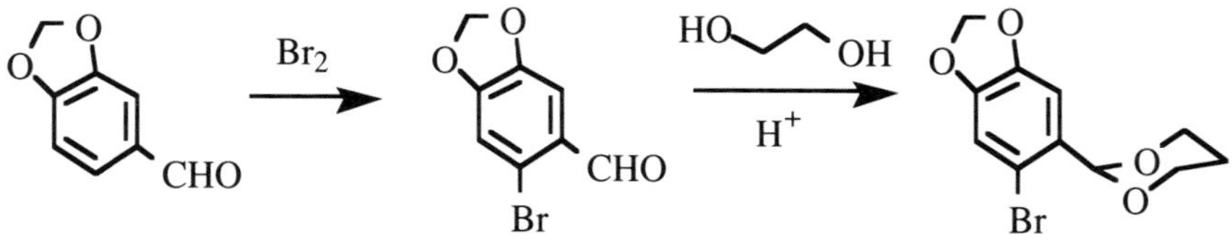

(**46**) (**47**)

The lower half of the biaryl is made by a slightly longer but straightforward route. Bromination of 3,4,5-trimethoxybenzoic acid gives

(**48**) which undergoes an Ullman-type reaction with methoxide in the presence of copper to afford the tetramethoxy compound (**49**). This is converted *via* the amide to the imidate (**50**), which reacts with the chiral auxiliary (**51**) (see section 5.3.2) to give the chiral oxazoline (**52**).

For the coupling of the two parts the Grignard reagent of (**47**) is prepared and reacted with (**52**). The crucial aryl–aryl bond is formed with a diastereomeric ratio of 7:1. The major diastereomer (**53**) is separated from the minor (**54**) by chromatography, and hydrolysed to aldehyde (**55**). At this point a salient feature of chiral biaryls reveals

itself, namely that racemisation can occur by rotation about the central bond. The aldehyde group, being sp^2 hybridised and therefore planar can slip past the ortho substituents on the other ring, allowing rotation to occur. For this reason the hydrolysis is performed at low temperature, and the aldehyde (**55**) is reacted immediately with methyl magnesium bromide. The resulting secondary alcohols are separated chromatographically and diastereomer (**56**) used for the next step. Both (**56**) and its diastereomer were totally stable to racemisation, even in boiling toluene, since they possess an sp^3 hybridised carbon in place of the aldehyde.

The secondary alcohol is protected as the allyl ether (**57**), and the chiral auxiliary, having served its purpose, is removed by acid hydrolysis to the aminoester salt (**58**), which is then reduced to the

primary alcohol (**59**). This in turn is converted into the benzylic bromide with triphenylphosphine and *N*-bromosuccinimide, and reacted with malonate anion, affording (**60**). The allyl ether is removed by isomerisation to the vinyl ether and hydrolysis with mercuric chloride, and the resulting alcohol (**61**) oxidised and α-brominated to the cyclisation precursor (**62**). Since there is now an sp^2

carbon atom in the ortho position, racemisation is again a danger so cyclisation with potassium t-butoxide to (**63**) is performed immediately.

At this point the synthesis intersects with an earlier racemic synthesis of Ziegler. Thus the diester is hydrolysed and decarboxylated, and the monoacid re-esterified with diazomethane, giving a 1:1 mixture of conformational isomers (**64**) and (**65**), which can be separated easily and re-equilibrated thermally. There is a rather subtle stereochemical point to be noticed here: despite the fact that rotation about the aryl–aryl bond occurs, there is no racemisation. The thermal interconversion does not occur through a planar transition state, due to the extreme conformational restraints on the cyclo-octanone bridge.

To complete the synthesis, the undesired isomer is recycled by thermal re-equilibration and chromatographic separation. Several cycles allow a ~90% conversion to (**64**). The remaining carbon atom is installed by an aldol reaction with formaldehyde in aqueous base, which occurs *anti* to the methoxycarbonyl group. Lactonisation occurs spontaneously. Under the conditions of the aldol reaction, the ketone is reduced in a crossed Cannizarro reaction with formaldehyde. This is corrected by re-oxidation with Jones reagent, affording the natural product (−)-steganone. Unfortunately, the last two steps proceed in an overall yield of only 11%. The authors discovered that their sample of steganone had an enantiomeric excess of 80–84%, judging by the optical rotation. It is their view that about 10% racemisation occurs at the α-bromination stage (**61**) → (**62**).

7.2.5 (+)-α-Allokainic acid (Oppolzer et al.[8])

This very short synthesis of the neurotoxin (+)-α-allokainic acid demonstrates how the understanding of the transition state of a reaction can be used to predict its stereochemical outcome. The chiral auxiliary is (−)-8-phenylmenthol (**66**), which is prepared from (+)-pulegone. Its Z-chloroacrylate ester (**67**) is reacted with diethyl trifluoroacetamidomalonate, giving (**68**). Note the retention of the Z-configuration of the double bond. This in turn is alkylated on nitrogen with prenyl bromide to give the cyclisation precursor (**69**). It is at this point that the reason for the choice of chiral auxiliary becomes apparent. The two faces of the double bond are non-equivalent by virtue of the chiral auxiliary (diastereotopic). The *Si* face (at position*) is encumbered by the bulky dimethylbenzyl group, and therefore the Lewis acid catalysed intramolecular 'ene' reaction takes place preferentially from the *Re* face. If 8-phenyl-menthol is replaced by menthol itself, the asymmetric induction is far lower, as one would expect. Hydrolysis and decarboxylation of the pyrrolidine (**70**) gives the natural product, in which the stereogenic centre adjacent to the nitrogen has equilibrated to the more stable configuration with the acid function *anti* to the neighbouring side-chain.

7.2.6 *Leiobunum* defence secretion (Enders[9])

Enders synthesises this simple compound, the defence secretion of the 'daddy longlegs' spider, *Leiobunum vittatum* and *L. calcar*, to illustrate the use of his RAMP and SAMP chiral auxiliaries (see section 5.3.2). The imine of 3-pentanone and SAMP is deprotonated with LDA, and the anion alkylated with *E*-1-bromo-2-methylbut-2-

ene. Once again, intramolecular chelation of lithium by the auxiliary is the key to the extremely high diastereofacial selection observed. Removal of the SAMP group gives the chiral (S)-ketone which has an enantiomeric excess of better than 95%. The (R)-enantiomer is made simply by substituting RAMP for SAMP. Though more conventional syntheses could be devised, they are unlikely to be as short and enantiospecific as this.

7.3 Third-generation methods

7.3.1 Carbocyclic iododeoxyuridine (Roberts et al [10])

Modified and unnatural nucleotides (e.g. AZT) are of great current interest as antiviral agents against diseases such as AIDS, as well as for antineoplastics. The target molecule, a carbocyclic analogue of the known agent 5-iododeoxyuridine, was synthesised in chiral form using the chiral hydroborating agent (−)-diisopinocampheylborane (**72**) in the crucial step (see section 6.4.3).

The substrate for the asymmetric reaction is the symmetrical cyclopentadiene (**71**), celebrated for its key role in Corey's early prostaglandin syntheses. Reaction of (**71**) with the bulky (−)-(Ipc)$_2$BH (**72**) results in diastereoselective attack *anti* to the benzyl-oxymethyl group but also enantioselectivity between the two enantiotopic double bonds. Oxidative work-up of the borane (**73**) gives unsaturated alcohol (**74**). Hydroxyl-directed epoxidation and benzylation gives epoxide (**75**), which opens regioselectively on treatment with the dianion of uracil, affording (**76**). The synthesis is completed by radical deoxygenation to remove the hydroxyl group, debenzylation and finally iodination of the heterocycle.

7.4 Fourth-generation methods

7.4.1 A tricyclic analogue of indacrinone (Dölling et al. [11])

The title compound is a uricosuric which is currently in clinical development by Merck.

"Carbocyclic IDU"

ASYMMETRIC TOTAL SYNTHESIS

This synthesis, which was reported by a group of development chemists, represents a remarkably efficient application of asymmetric alkylation by chiral phase transfer catalysis (PTC) (see section 6.1.1). Reaction of indanone (**77**) and allylic halide (**78**) under PTC conditions in the presence of only a few per cent of chiral cinchonidine derivative (**79**) gives in excellent yield and enantiomeric excess the (*R*) product (**80**).

220   ASYMMETRIC SYNTHESIS

The high enantioselectivity is ascribed to preferential ion-pairing on one enantiotopic face of the enolate of (**77**) with the chiral quaternary ammonium ion of the catalyst. The remaining steps consist of acid hydrolysis of the vinyl chloride and cyclisation to (**81**) (the Wichterle reaction) and installation of the acetic acid side-chain. It is likely that this asymmetric synthesis is actually used for large-scale production of the product.

7.4.2 (S)-(−)-Propranolol (Sharpless et al.[12])

Perhaps the most important recent development in the field of asymmetric synthesis has been the famous Sharpless epoxidation. It is appropriate that this chapter includes the asymmetric synthesis of the beta-adrenergic blocker (S)-(−)-propranolol, reported by Sharpless himself.

ASYMMETRIC TOTAL SYNTHESIS

The key chiral intermediate is of course (*S*)-glycidol (**82**) prepared from allyl alcohol by the recent catalytic variant of the Sharpless epoxidation, using (+)-diisopropyl tartrate as the source of asymmetry. Glycidol is water-soluble and difficult to isolate, and so was reacted *in situ* with 1-naphthoxide anion to give (**83**). The authors point out a rather subtle and unobvious fact about chiral glycidol: the Payne rearrangement does **not** invert the stereogenic centre. Under the reaction conditions this base-catalysed rearrangement is undoubtedly occurring, but does not affect the outcome:

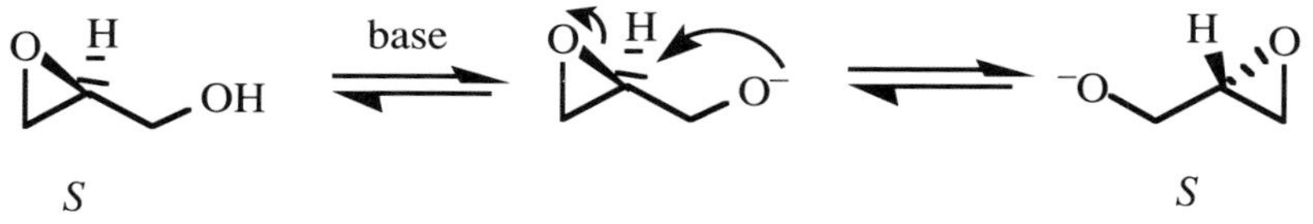

The synthesis continues by conversion of (**83**) to epoxide (**84**) and nucleophilic epoxide opening at the terminus with isopropylamine, giving (*S*)-propranolol of 98% e.e.

The enantiomer of (**82**) could also be used in the synthesis of (*S*)-propranolol. Attack of the phenoxide on the tosylate of (**85**) occurs almost entirely, but not exclusively, by direct nucleophilic displacement rather than epoxide opening followed by displacement.

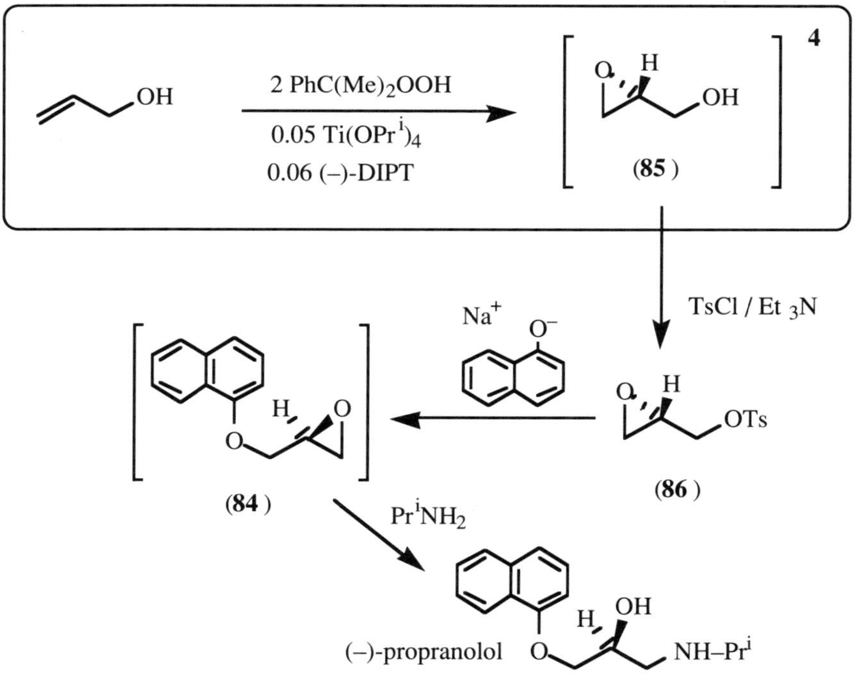

This route is even shorter than the first. As before, (*R*)-glycidol (**85**) was not isolated, but instead converted directly to tosylate (**86**). This, on reaction with the phenoxide, gave a 97:3 mixture of (**84**) and its enantiomer, which results from attack on the epoxide. The final step was performed without isolation of (**84**). Recrystallisation of the propranolol removed the small proportion of the undesired (*R*)-enantiomer, giving material which was essentially enantiomerically pure. Perhaps the most impressive feature of this synthesis is the fact than in only two flasks allyl alcohol is converted on a 20 g scale to an enantiomerically pure product.

7.4.3 Takasago (–)-menthol synthesis[13]

Natural (–)-menthol is synthesised commercially on a multi-ton scale from the inexpensive achiral precursor myrcene in an elegant application of the asymmetric 1,3-hydrogen shift (see section 6.4.2). The synthesis is so good that it competes with the traditional isolation method from oil of wintergreen.

Myrcene is treated with lithium diethylamide, whereupon the amide adds to the diene giving the Z-allylamine (**87**). Asymmetric isomerisation with the rhodium complex of (*S*)-BINAP gives the enamine (**88**) in almost 100% enantiomeric purity and chemical yield. The enamine is then hydrolysed. Treatment of the resulting aldehyde (**89**) with the Lewis acid $ZnBr_2$ results in an intramolecular ene reaction, giving a cyclic product (**90**) (note the chair-shaped transition state, and how the remote equatorial methyl group controls the relative, and thus absolute stereochemistry of the two new stereogenic centres). Finally, the double bond is hydrogenated, giving (–)-menthol. As a bonus, a hydroxyaldehyde (**91**) with a lily-of-the-valley scent is obtained by hydration of the double bond in the aldehyde (**89**).

7.4.4 (–)-Prostaglandin E_1 (Sih et al. [14])

In this short synthesis of (–)-PGE_1 two of the four stereogenic centres are installed by microbial reduction. The substrate for the first asymmetric step is cyclopentanetrione (**92**), which is prepared in one step by a double Dieckmann condensation.

Incubation of (**92**) with the micro-organism *Dipodascus uninucleatus* gives a very good yield of *enantiomerically pure* 4*R*-hydroxydione (**93**). Interestingly, if the reaction is performed with *Mucor rammanianus* the opposite enantiomer is obtained. Regioselective *O*-pivaloylation, 1,2-reduction of the remaining carbonyl and mild acid hydrolysis accompanied by dehydration afforded (**94**), which was also made by Stork, but in eleven steps (section 7.2.1).

The stereogenic centre in the side-chain was formed by reduction of iodoenone (**96**) with *Penicillium decumbens*. The product (**97**) which was shown to have an e.e. of 80%, was converted to cuprate (**98**) and reacted with cyclopentenone (**95**). The hydroxyl protecting groups were removed in the work-up, giving the methyl ester (**99**) of (−)-PGE$_1$. The final step, hydrolysis of the ester, was also performed microbially, with *Rhizopus oryzae*.

7.4.5 Compactin and mevinolin analogues (Johnson and Senanayake[15])

The title compounds are natural products of microbial origin with potent inhibitory activity against the enzyme 3-hydroxy-3-methylglutaryl CoA reductase, which mediates a key step in human cholesterol biosynthesis. Thus they constitute a conceptually new and exciting approach to the prevention of the atheromatous plaques of ischaemic heart disease and arteriosclerosis.

This synthesis of the lactone portion illustrates how an esterase enzyme can distinguish between two enantiotopic groups in an achiral *meso* compound to provide an enantiomerically enriched chiral product.

R = H Compactin
R = Me Mevinolin

(100) → i. NaBH$_4$; ii. Bu^{t}Me$_2$SiCl / imidazole → (101)

(101) → $^1O_2$ → (102)

(102) → Zn or SmI$_2$ → (103)

(103) → i. Ac$_2$O / Et$_3$N / 4-DMAP; ii. HF → (104)

(104) → *Electric eel cholinesterase* → (105) **4**

The starting material for the synthesis is tropone (**100**). Sodium borohydride reduction of (**100**) proceeds by 1,8-addition of hydride to the trienone system, followed by kinetic reprotonation and then normal ketone reduction to give (**101**) after hydroxyl protection. The *cis* diene system of (**101**) reacts stereoselectively with singlet oxygen to afford a 5:1 mixture of endoperoxide (**102**) and its diastereomer, which are separated by chromatography. The peroxide bond is reduced to give (**103**) and this is acetylated to give the symetrical *meso* diacetate (**104**) which is the substrate for the crucial asymmetric step. On treatment with electric eel cholinesterase, only the pro-*S* acetate is hydrolysed to give (**105**) with an e.e. of better than 95%, and a chemical yield of 79%. An indication of the efficiency of this esterase can be gained from the fact that only 1 mg was used per 1 g of (**104**). With the two allylic oxygen atoms now differentiated, the synthesis can diverge to produce either enantiomer of the final products.

For the natural series, (**105**) is oxidised stereospecifically at the allylic hydroxyl with MnO_2 and then silylated at the remaining secondary alcohol to give enone (**106**). This compound, on treatment with diisobutylaluminium hydride and catalytic methyl copper, undergoes 1,4-reduction. The resulting aluminium enolate is transmetallated to the lithium enolate with methyllithium, presumably to make it more reactive, and then silylated. Ozonolysis of the resulting silyl enol ether (**107**) with reductive workup and diazomethane treatment affords hydroxyester (**108**). To complete the synthesis, a 'dummy' octalin in the form of a phenyl group is introduced by phenyl cuprate displacement of the tosyl group of (**108**). Double deprotection and lactonisation with HF gives the compactin/mevinolin analogue (**109**). Synthesis of the opposite enantiomer (**110**) follows the scheme shown after inversion of the pattern of protection.

References

1. P. Magnus and G. Roy, *J. Chem. Soc. Chem. Commun.*, 1978, 297.
2. G. Stork and T. Takahashi, *J. Am. Chem. Soc.*, 1977, **99**, 1275.
3. P.J. Reider and E.J.J. Grabowski, *Tetrahedron Lett.*, 1982, **23**, 2293
4. D.A. Evans and J. Bartroli, *Tetrahedron Lett.*, 1982, **23**, 807.
5. K.A. Nelson and E.A. Mash, *J. Org. Chem.*, 1986, **51**, 2721.
6. R.C. Andrews, S.J. Teague and A.I. Meyers, *J. Am. Chem. Soc.*, 1988, **110**, 7854.
7. A.I. Meyers, J.R. Flisak and R.A. Aitken, *J. Am. Chem. Soc.*, 1987, **109**, 5446.

ASYMMETRIC SYNTHESIS

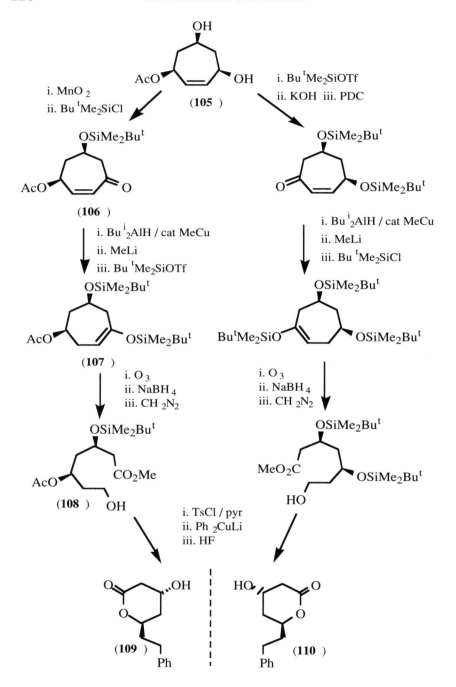

8. W. Oppolzer, C. Robbiani and K. Bättig, *Helv. Chim. Acta*, 1980, **63**, 2015.
9. D. Enders, in *Current Trends in Organic Synthesis*, H. Nozaki, ed., Pergamon, Oxford, 1983, p.151.
10. K. Biggadike, A.D. Borthwick, D. Evans, A.M. Exall, B.E. Kirk, S.M. Roberts, L. Stephenson, P. Youds, A.M.Z. Slawin and D.J. Williams, *J. Chem. Soc. Chem. Commun.*, 1987, *251*.
11. A. Bhattacharya, U.-H. Dölling, E.J.J. Grabowski, S. Karady, K.M. Ryan and L.M. Weinstock, *Angew. Chem., Int. Ed. Engl.*, 1986, **25**, 476.
12. J.M. Klunder, S.Y. Ko and K.B. Sharpless, *J. Org. Chem.*, 1986, **51**, 371.
13. S. Otsuka and K. Tani, in *Asymmetric Synthesis*, Vol. 5, J.D. Morrison, ed., Academic Press, Orlando, 1985, Chapter 6.
14. C.J. Sih, J.B. Heather, R. Sood, P. Price, G. Perruzotti, L.F. Hsu Lee and S.S. Lee, *J. Am. Chem. Soc.*, 1975, **97**, 865.
15. C.R. Johnson and C.H. Senanayake, *J. Org. Chem.*, 1989, **54**, 735.

Index

absolute configuration
  determination   20
  specification   22
alkaloids as starting materials   67
allokainic acid   213
alpine borane®   176
amino acids
  as starting materials   64
  synthesis of   90, 106, 108
amino alcohols as starting materials   64
ancistrocladine   17
ant alarm pheromone   103
aspartame   4
asymmetric reactions
  α-alkylation of acids   104, 108
  α-alkylation of amines   118
  α-alkylation of δ-ketoacids   104
  α-alkylation of ketones   103, 143, 217
  α-alkylation of sulphoxides   119
  aldol reaction   109
  alkylation, catalytic   143
  [2+2] cycloaddition   150
  Claisen-Cope rearrangement   139
  cyclopropanation   89, 203
  Diels-Alder cycloaddition   135, 152
  dihydroxylation of alkenes   163
  dihydroxylation of arenes   187
  double bond isomerisation   173, 222
  epoxidation   160, 165, 220
  formation of alkenes   120, 155
  α-hydroxylation of acids   108, 167
  α-hydroxylation of ketones   165
  hydroboration   174
  hydrogenation of alkenes   167
  Michael reaction   104, 123, 144
  nucleophilic addition to carbonyl   132, 149
  osmylation   163
  oxidation of *meso* diols   185
  oxidation of sulphides   162
  reduction of carbonyl compounds   132, 172, 176, 179, 181, 223, 224
  Robinson annulation   159
asymmetric synthesis
  auxiliary controlled, definition   74

  auxiliary controlled, examples   102
  auxiliary controlled, general   102
  catalyst controlled, definition   76
  catalyst controlled, examples   144
  definition   5
  first-generation, definition   73
  first-generation, examples   92
  fourth-generation, definition   76
  fourth-generation, examples   143
  mechanism   80
  reagent controlled, definition   75
  reagent controlled, examples   143
  second-generation, definition   74
  second-generation, examples   102
  second-generation, general   102
  substrate controlled, definition   73
  substrate controlled, examples   93
  third-generation, definition   75
  third-generation, examples   143
asymmetric Wittig reaction   156
atropisomers   15
avenaciolide   93
aztreonam   91

baker's yeast   182
BINAP   168, 222

Cahn–Ingold–Prelog (CIP) system   22
CAMP   11
captopril   109
carbocyclic iododeoxyuridine   216
carbohydrates as starting materials   69
chiral acetals   130, 203
chiral acids, protonation by   156
chiral alkylidenecycloalkanes   14
  synthesis   120
chiral allenes   14
chiral azaenolates   102
chiral bases   158
chiral biaryls   15
  synthesis   125, 206
chiral borohydrides   177
chiral building block approach   70
chiral camphor sultams   128, 136
chiral derivatising agents (CDAs) for NMR   44

chiral diketopiperazines 106
chiral eluant HPLC 42
chiral enolates 102
chiral formamidines 117
chiral hydrazones 104, 215
chiral iron acyls 108
chiral lanthanide shift reagents (CLSRs) for NMR 44
chiral nitrogen compounds 10
chiral oxathianes 133
chiral oxaziridines 165
chiral oxazolidinones 108, 112, 136, 200
chiral oxazolines 104, 123, 203, 204
chiral phosphorus compounds 11
chiral silicon compounds 10
chiral solvating agents (CSAs) for NMR 50
chiral stationary phase
 GC 36
 HPLC 40
chiral sulphoxides 119, 126, 132
 synthesis 162
chiral sulphoximines 87
chiral sulphur compounds 12
chirality 1
 axial 13
 conformational 15
 due to isotopes 10
 transfer of 31
 types of 8
chirally modified $LiAlH_4$ 179
cinchonidine 17, 219
cinchonine 17, 144
circular dichroism (CD) 36
citronellol 97
compactin 224
Cram model ('rule') 101
crotylboranes 153
cypermethrin 84

daddy longlegs defence secretion 215
deltamethrin 5
deracemisation 157
dextrorphan 3
diastereomer ratio (d.r.) 18
diastereomeric excess (d.e.), definition 18
diastereomers, definition 15
diastereoselectivity, definition 18
diastereospecific, definition 18
diastereotopic, definition 20
DIPAMP 11, 169
DOPA 3, 169
double asymmetric induction 79, 116, 202

enantiomeric excess (e.e.)
 definition 6
 determination 33
enantiomeric purity of natural products 7
enantiomerically pure compounds 6

enantiomers, definition 1
enantioselectivity, definition 6
enantiospecific, definition 6
enantiotopic, definition 19
enzymes 181
esterases 188, 227

frontalin 192

gas chromatography, chiral column 36

hamatine 17
HLADH 185
HPLC
 chiral column 39
 chiral eluant 42
HR 780 98
hydroxy acids
 as starting materials 66
 synthesis of 108, 150

indacrinone 143
 analogue 218
Ipc_2BH 174, 217
$IpcBH_2$ 176

kinetic resolution 77, 161

l/u nomenclature 25
lk/ul nomenclature 28
lactic acid 100
latrunculin B 85
levorphanol 3
lipases 189

mannitol 70, 97, 193
menthol, asymmetric synthesis 222
meso compounds 27, 185
mevinolin 224
milbemycin β_3 97
Mosher acid (MTPA) derivatives 45
muscone 203

NMR
 using chiral derivatising agents 44
 using chiral lanthanide shift reagents 44
 using chiral solvating agents 50
nomenclature, summary 32

optical activity 2, 33
optical purity 35
optical rotation, measurement 33
optical rotatory dispersion (ORD) 36
Oriental hornet pheromone 120

INDEX

oxetanocins 152

permethric acid 83
phosphonothricin 107
PLE 188
podophyllotoxin 203
podorhizon 126
polarimetry 33
Prelog–Djerassi lactone 200
prochilarity 19
propranolol 3, 200
prostaglandins 186, 193, 222
PS 5 100
pseudomonas 187

quaternary centres, asymmetric synthesis of 106
quinidine 17, 150, 163
quinine 17, 145, 150, 163

$R*R*/R*S*$ nomenclature 25
R/S nomenclature 23
racemate, definition 6
RAMP 103
Re/Si nomenclature 27

relative configuration, specification 25
relative topicity of reactions, specification 28
resolution 71
    examples 83

SAMP 103, 215
self regeneration of stereogenic centres 127
Sharpless dihydroxylation 163
Sharpless epoxidation 160, 200
steganone 206
stereogenic centres 4

terpenes as starting materials 68
thalidomide 2
thienamycin 196
thujopsene 90
topicity of reactions, specification 27
triple asymmetric induction 79

Wieland–Miescher ketone 159

yohimbone 119

Zimmerman–Traxler model 110